Yale Agrarian Studies Series

JAMES C. SCOTT, SERIES EDITOR

The Agrarian Studies Series at Yale University Press seeks to publish outstanding and original interdisciplinary work on agriculture and rural society—for any period, in any location. Works of daring that question existing paradigms and fill abstract categories with the lived-experience of rural people are especially encouraged.

For a complete list of titles in the Yale Agrarian series, visit www.yalebooks.com

James C. Scott
Series Editor

Squeezed

What You Don't Know About Orange Juice

ALISSA HAMILTON

Yale University Press   New Haven and London

Published with assistance from the Mary Cady Tew Memorial
Fund

Printed in the United States of America.

The Library of Congress has cataloged the hardcover edition
as follows:

Hamilton, Alissa.
Squeezed: what you don't know about orange juice / Alissa
Hamilton.
    p.   cm—(Yale agrarian studies series)
  Includes bibliographical references and index.
   ISBN 978-0-300-12471-2 (hardcover: alk. paper)
1. Orange juice industry—Florida—History. 2. Orange juice—
Florida—History. I. Title. II. Title: What you don't know
about orange juice. III. Series: Yale agrarian studies.
HD9348.5.O723H35 2009
338.4'766363—dc22

                                     2008049893

ISBN 978-0-300-16455-8 (pbk.)

A catalogue record for this book is available from the British
Library.

10 9 8 7 6 5 4 3 2 1

For my mother
Her spirit is in this

Contents

Part II—Developing Orange Juice Standards of Identity

Part III—Florida's Orange Juice Industry Post-1960

Preface

While sitting at my computer one afternoon, I heard a wine connoisseur on the radio. She said that the diversity and complexity of wines and their flavor explain why we need wine experts and not orange juice experts. The comment roused a "not true" from me, but not surprise. In the popular imagination orange juice has come to symbolize the opposite of complexity—simplicity defined.

The radio guest is not alone in assuming the orange juice she pours each morning is straightforward. Informal chats with colleagues, friends, family, and strangers reinforce that the majority of orange juice drinkers believe their favorite brand of orange juice is "simply orange." Few are aware of the countless men and women in white lab coats behind every carton of "pure," "100 percent" orange juice. Whether you buy "not from concentrate" or "from concentrate" orange juice, you can be sure part of the cost pays for the work of scientists who spend their lives attempting to mimic the highly elusive flavor of freshly squeezed orange juice.

Despite the multiple layers that make up orange juice, the first question most ask when they hear about a book on orange juice is, "Is orange juice good for me?" Next is: "Should I drink it? And, if so, what kind?" And then there is the silent question that the raised eyebrows imply: "A *book* on orange juice?"

This book is not about whether orange juice is healthy or whether you should buy it. Orange juice has a lot to say about a lot of issues other than whether it is a good source of vitamin C. It has something to say about trade policy and the growing phenomenon of foreign food dependency; orange pickers, who are composed of mostly migrant labor and who toil often under horrendous conditions behind the pastoral scenes that orange juice companies paint on cartons and billboards and show on television; Anita Bryant and how she became, to those of a certain age, synonymous with orange juice; and more.

This book is primarily a history of Florida's orange juice industry and the stories that its most vocal members, the brand labels, tell. Florida orange growers are central to the plot. The book is also about government regulators who are supposed to pay attention to the industry's messages, and consumers, who, orange juice consumption statistics show, do. Companies with a product to sell deliver these messages mainly for one reason: to attract consumers. Unfortunately, industry-manufactured messages can be as unhealthy as industry-manufactured food.

So why orange juice? Because almost everybody drinks it and is misinformed about what it is they are drinking. I wrote the book with a modest ambition: to make you look at your glass of presqueezed orange juice differently and begin to see through the opaque packages of food that surround you. Most look the other way. This book encourages you to open your eyes to the hidden and strange ways of our modern, industrialized, and global food environment.

Acknowledgments

Where to begin and end. From kindergarten through high school my parents were always on the lookout for teachers and schools that would teach me how to love to learn. They succeeded.

Thank you big sister. As my imagined reader you kept me company and on my toes while writing.

Big thanks to John Wargo, Daniel Kevles, and Jim Scott. John provided the balance of support and freedom that I needed to write something more than a dissertation. I could always rely on Daniel for thorough readings and critical, in both senses of the word, comments. Jim helped me step back and see the bigger picture. Thank you also to Benjamin Cashore, who offered valuable suggestions regarding the book publishing process.

Thank you Elisabeth Barsa for your early interest. I am pretty sure you were the first to read anything I wrote on orange juice. You gave me hope that somebody other than my committee would want to read what I was writing.

I am so grateful to those I interviewed. I spoke with a wide range of members of the orange juice industry while spending the spring of 2004 in Florida researching. Special thanks to Edwin Moore, inventor of frozen concentrated orange juice, Robert Barber, Robert Braddock, Jim Brewer, Mark Brown, Bill Castle, Rolland Dilley, Kevin Gaffney, Sharon Garrett, Fred Gmitter, Renée Goodrich, Jim Griffiths, Jude Grosser, Charles Hendrix, Michael Kesinger, Daniel King, Orie Lee, the Martin family, Allen Morris, Ronald Muraro, Herbert Nigg, Philip Rucks, Mike Sparks, Thomas Spreen, and those at Tropicana and Firmenich whose identities shall remain anonymous. All were extremely generous with their time. Each contributed a unique perspective on, and insights into, the industry.

Thanks to the late Thomas Mack, Donna Logan, and everyone who donated to the Thomas B. Mack Citrus Archives in Lakeland, Florida. Donna and Professor Mack were extraordinarily patient with my inquiries and exhaustive searches of the archives. I found many valuable documents there, including the hearings into developing standards of identity for orange juice that are central to the book and that no library I know of holds. Thanks to all at Florida's Citrus Research and Education Center, especially librarian Pamela Russ, who helped me find some obscure material. Thank you also to professors and staff at Florida Southern College. Everyone there was very welcoming.

Thank you to Jean Thomson Black, Executive Editor at Yale University Press, for seeing the book through to publication. Her vision and her encouragement from day one has given me the opportunity to thank her here. Thanks also to Matthew Laird for efficiently pushing the manuscript into editing and production. A warm thank you to Jeff Schier, who has made your read cleaner and more enjoyable. He is a dream copyeditor, the kind who asks the important questions and makes changes only for the better while respecting the author's voice and intent. The downside is that I therefore must take full responsibility for any awkward or difficult moments.

I would like to acknowledge the Food and Society Policy Fellows program and the Woodcock Foundation, which is funding me as a Fellow. The program is full of encouraging and inspiring people working to make healthy, green, fair, and affordable foods available to all. Thanks especially to Judith Weinraub for challenging, and thereby helping me to clarify, my thinking and writing about the role of a consumer right to know how food is produced.

Finally, and importantly, my deepest gratitude to the great problem solvers Marilyn Zivian, who is right on *all* the time, and Ghada Sharkawy, who is wise beyond her years. Both continue to provide unwavering support. True friends.

Funding for research was provided by Yale University's graduate school; Yale's Department of Forestry and Environmental Studies; the Heinz Family Foundation; Homegrown Naturals Inc.; the John Perry Miller Fund; and Yale's Institute for Biospheric Studies.

Thank you all.

Introduction

My original plan for this book was to investigate the orange juice industry's effect on the biological diversity of the sweet orange, but the changing genetic makeup of the Florida orange became peripheral to my research almost as soon as I arrived in Florida. After a few interviews I learned that the juice industry alone has not significantly affected sweet orange diversity. And then I found a needle in a haystack that threaded together bigger and more compelling questions to investigate. The journey to the pinpoint discovery had a few detours.

In February 2004 Dixi, my nine-pound Jack Russell–Chihuahua companion, and I packed up for a long drive from Toronto, Canada, to Lakeland, Florida. Located in the heart of orange growing and processing territory, the small dot on the center of the map of Florida proved to be the ideal base camp. And camp it was. The unheated, un-air-conditioned, cockroach-, termite-, and flea-infested coach house that Dixi and I lived in for four months offered little protection against Florida's unpredictable climate and thriving bug life.

Lakeland's relative proximity to the bulk of the state's orange juice processors and growers made it an obvious choice to park for a few months. The small town, home to Florida Southern College, also boasted many unexpected advantages. For one, Florida Southern has a strong citrus department. Members of the faculty pointed me to key players in the industry. Another bonus was the Garden-of-Eden-like campus, resplendent with "golden apple" trees growing in dense patches. Waking each morning to the heavy fragrance of orange blossoms I began to understand why so many Floridians are steadfast that the forbidden fruit Eve offered Adam was a golden apple, or, in more common parlance, an orange.

Florida Southern grows more than oranges. The campus is planted to more types and varieties of citrus than most people know exist, and produces some of the best grapefruits, oranges, and kumquats I have ever tasted. Store-bought does not compare. Strangely the students seemed oblivious to the riches fruiting around them, more unreal to me than money growing on trees. Better, too; money does not taste so sweet. Although the citrus department that manages the trees encouraged picking, I never saw anybody twisting the fruit, which loosens beckoningly when ripe, off the trees. An innate distrust of anything not grown for sale, perhaps. Ignoring the taboo, I picked as much fruit as I could eat in a day, never tiring of the real thing.

Plentiful and always fresh citrus was a treat. But the biggest of Lakeland's surprises was the mint-colored house down the street. Behind the Disney-like façade of the two-storied house were the Thomas B. Mack Citrus Archives. I had heard in passing that Lakeland had something called a citrus archives. I had no idea what an untapped gold mine of orange juice information it would be. The eponymous Archives were essentially one man's collection of everything citrus. In his early nineties at the time, Professor Mack, as everyone knew him, had been building the collection for over half a century.

Professor Mack, who died soon after I left Florida in late spring 2004, was one of those characters any historian is lucky to meet. After years teaching at Florida Southern College, he turned his energies to organizing and maintaining the Archives, for which Florida Southern provided space and funding. Donna the archivist was his only help. He rarely missed a day at the office, a small room on the first floor of the converted residential house.

At least once a day Professor Mack could be found on the front porch in his rocking chair. Donna would move the rocking chair outside each morning when she arrived, and back inside before closing up for the evening, never sure when her frail yet spirited boss would decide to take a breath of fresh air. The rocking chair elicited hours of reminiscing from Professor Mack about the days when the orange juice industry was just beginning. He had watched the industry germinate into a giant, collecting weird and wonderful memorabilia along the way. I remain indebted to him and his rocking chair, at one time permanent fixtures of 901 Callahan Court, for hours of recorded Mack memories and for my memory of the picturesque scene of the very un-twenty-first-century routine.

Professor Mack's second home, the Archives, became the same for Dixi and me. He and Donna welcomed us with open arms—literally in the case of Dixi. In the unusual event Dixi could not accompany me on one of my many sojourns, she stayed at the Archives on Professor Mack's lap or by Donna's side. Other visitors were rare. Over the course of the many months I frequented the Archives, I crossed paths with only one user, a graduate student. Interested in the Archives' collection of old orange-related newspaper articles, he quietly went about his work for a day or two. Dixi, protecting her new family and territory, barked every time he entered. The intruder did not seem to mind.

When Professor Mack was not taking me back with him to central Florida of the 1940s, '50s, and '60s, he and Donna gave me free rein to rummage through the Archives' boxes, file cabinets, and shelves full of mostly yet-to-be indexed papers. During one unmethodical search, the only type possible among the disarray, I came across a blue-covered spiral bound document that was lying inconspicuously in an unnamed filing cabinet. I pulled it out and discovered from its title that it was one of twenty-seven volumes. A little more searching amazingly produced the complete set, 3,420 pages in total. The front cover bore the stamp of the U.S. Food and Drug Administration (FDA). Black print declared that the "matter" was "Orange Juice and Orange Juice Products; Definitions and Standards of Identity." I soon learned that a "standard of identity" is an FDA creation for regulating food, staples mostly. Although still in the dark as I stood over the filing cabinets in which the softbound volumes, interspersed with unrelated material, slouched, I sensed their importance.

Once I sat down and flipped through Volume One I realized I had before me a critical record of orange juice history: FDA hearings into establishing the equivalent of government enforceable recipes for the making and marketing of the various orange juice products on the market. I gathered from the first few pages that until the hearings, which began in January 1961, commercial orange juice was relatively unregulated. Disparities in quality among similarly labeled juices, for instance some "canned orange juice" contained undeclared sweeteners while others did not, prompted the FDA to standardize the products.

What was most intriguing about the document was not the FDA's decision to regulate orange juice. Rather, bound up inside were thousands of pages of opinions, thoughts, and facts about orange juice, compiled at a critical moment in the orange juice industry's history, when it was growing exponentially. The document's timing and inside story made it an irresistible read. There had to be a reason orange juice attracted so much attention. I was eager to find out why. I delved in immediately, pacing myself with a minimum of two volumes per day, taking copious notes along the way.

On the first page I was struck to learn that counsel for orange juice processors, and their partners in crime, sometimes literally, dominated the list of those present. Tropicana Products, Minute Maid, Sunkist Growers, various smaller orange juice processors, the National Orange Juice Association, the National Association of Frozen Food Packers, the U.S. Beet Sugar Association, the American Molasses Company, and the National Dairy Products Corporation—all had legal representatives attending. The list betrayed the FDA's heavy reliance for guidance on the same members of industry whose habits provoked FDA action.

Add to the mix of those present at the hearings counsel for the FDA, state boards of health, and a few consumer organizations, and then factor in the argumentative nature of the hearings' trial-like procedure, and it is no wonder that the gathering, symbolically "Before the Secretary, Department of Health, Education and Welfare, Food and Drug Administration," but in fact before chief Examiner William Risteau, often turned combative.

The testimony of processors, food scientists, FDA officials, and a handful of consumer representatives spanned a wide range of subjects: from acceptable marketing and processing technology to consumer orange juice perceptions and expectations, to industry shenanigans. Through witness examinations and cross-examinations, counsel for the FDA and industry hashed out in painstaking detail the various topics discussed. The FDA had identified some of the issues beforehand. Others took the government officials who knew little about juice processing going in, and me, by surprise. Clearly there was a lot to say about orange juice—six months' worth of testimony.

In investigative procedures, hearings especially, you never know where a gem, whether a hidden fact, an unexpected confession, or an interesting digression, might turn up. I took advantage of the benefit I had as a reader after the fact. I processed every sentence, including the odd one that might have gone unheard by those attending, whether because muttered by the speaker or not listened to by the daydreamer. Time was on my side to decipher convoluted questions and answers that were likely not intended to be analyzed too closely by those present. The task was never boring. The script read as if written for theater. It contained multiple subplots, lead characters with whom I became familiar, and amusing, even witty dialogue.

What I did not know at the time was how lucky I was to have found a copy of the weighty manuscript. Months passed before I realized how limited its distribution is. I was alerted to the fact while interviewing an official at Tropicana. When I asked him a question about the hearings he looked at me blankly. He inquired about them on more than one occasion, asking first for a citation, then where he could find a hard copy. The more people I interviewed in Florida the more I understood how few had ever heard about, let alone read, the document, Professor Mack included. Despite his incredible memory, he could not remember how he acquired a copy.

The transcript's publication, by Columbia Reporting Company, gave me confidence that I would be able to find another copy when I returned to New Haven to write up my research. I was mistaken. The talented Yale librarians could not produce another copy of the elusive hearings. Not even the librarian at the FDA's history office had seen a copy, or knew where I could find one. In an e-mail the Yale librarian helping with the search summarized the situation by noting that the FDA History Office "may or may not have the exact same record . . . some of their holdings from that time period have been transferred to the National Archives, some are still at the FDA, some are at the Federal Records Center in Suitland, Maryland."

Her e-mail said I could file a Freedom of Information request with the caveat that "there are limits to how much photocopying they could do per day." As if to dispel any remaining hope, the e-mail ruled out interlibrary loan, which "is not an option from the FDA." The concluding words were not reassuring: "Good luck getting your hands on this nearly-unique resource." I filed away the e-mail certain of only one thing: the prospects of finding another copy of the hearings were not promising.

Although an obsessive thoroughness when it comes to note-taking slows the research process down, it pays off every now and then. While in Florida I amassed hundreds of my own, and photocopied pages to turn to when needed. Fortunately having the entire document in front of me was a luxury I could do without.

In June 2006 I received a phone call from a Lakeland librarian who said that the Thomas Mack Archives were being dismantled. Professor Mack had passed away, and Florida Southern no longer wanted to house the collection. The librarian had heard that I had used the Archives extensively and been advised to ask me what to do with them. The Archives were a mess. I honestly did not know how to divvy them up logically. I did know there was one very important twenty-seven-volume document sitting somewhere, probably not in the same place that I had found it, which should be dealt with carefully. But try to communicate this to one who has a problem on his hands and is looking for a quick and easy solution. I could sense that the librarian did not have time to go looking for the needle that I had dutifully put back in the haystack, so I did not try to send him on a search. To this day I do not know what happened to the pages upon pages of testimony regarding the inner workings of the orange juice industry, and the 1960s consumer to whom the industry catered.

There is a man at Tropicana who wants to know what the hearings are all about. For one they are about why Tropicana, now one of the most popular orange juice brands, was one of the FDA's most wanted. That is just the beginning of what they show and tell about the orange juice industry. Maybe you are not shocked to hear that deception is a major theme. The food industry has been in the news a lot recently. It has been widely criticized for its schemes to encourage consumers to eat calorie laden, nutrient deficient foods. The tactics have ranged from advertising sugar-, salt-, and fat-laden foods to children, to increasing portion sizes via packaging and "super-sizing," a common and economically attractive option in fast food restaurants until the movie Super Size Me demonized it. Given all the news about the processed food industry's subterfuges, you may have already concluded that all processed foods, including orange juice, are better left unconsumed.

Then again, maybe seeing orange juice and deception in the same sentence surprises you. Orange juice has come to symbolize purity in a glass. This book is partly about how orange juice marketers have succeeded in creating an aura of golden goodness around the product. The idea that orange juice is "an essential part of a balanced breakfast" is familiar and for the most part unchallenged. Criticism of the processed food industry focuses on that segment of the industry that produces and markets "junk" food, notably candy, chips, soft drinks, sugary breakfast products, and typical fast food fare. Little attention is paid to processed foods such as orange juice that are advertised as pure, fresh, and additive-free. From the informal interviews I have conducted, it is apparent that those who buy orange juice buy the stories that the industry tells, too. They are taken aback to hear that orange juice may not be what it is made out to be.

You may be wondering how much it all matters. Whether or not the industry's claims about orange juice are accurate, chances are the juice will not kill you. There are many products on the market that will. So why write a book about a relatively benign product? The answer is precisely because orange juice is not lethal. The history of processed orange juice and its marketing highlights the fact that as a society we tend not to care too much about deceptive advertising unless the product being pushed is measurably harmful. The tobacco industry came under fire for massive marketing campaigns that successfully promoted its addictive product. The negative press is understandable given that death has been shown to be a side-effect of buying the industry's former, now banned, slogans. Similarly, government officials, not-for-profit organizations, and academics are scrutinizing the ways that marketers attempt to increase junk food sales. Such action is equally understandable given the ploys to promote these foods have been linked to obesity, a national killer.

Then there is orange juice. Nobody hears or says much about the ads that have made orange juice into a beverage that unites North America across class, cultural, and geographic lines. Widespread ignorance about the difference between how orange juice is produced and promoted is partly to blame. But such ignorance is the symptom of a general lack of will to investigate and ultimately crack down on orange juice advertising. Orange juice advertisers still get away with messages that the FDA has repeatedly frowned upon. The implicit reason is that resources are scarce and must be used judiciously. For the FDA, which is responsible for food *and* drug regulation, new drug approval is naturally a priority. The advertising of a product that poses little if any health risk to the consumer is low on everybody's list.

But maybe it should not be. Although the tales that the orange juice industry spins do not kill, they catch. If you have ever watched a fly in a web, you have witnessed how swiftly and effectively entrapment maims action. Just as a spider's web restricts free movement, a deceptive advertisement, when carefully crafted, can restrict free thought. This is a big assertion. It leads to the conclusion that advertising poses a threat to a free, in the full sense of the word, society. Ask yourself why, like most people, you drink orange juice. You probably say the reason is that it is good for you, or that it is high in vitamin C, or that you grew up drinking it and like it. If so, then I must frankly tell you that, when it comes to orange juice, you are acting like a robot. If you are not convinced keep in mind, as you read on, how many products you buy and the reasons why.

No doubt orange juice is not the only product you go to the store and pick up thoughtlessly at best, and based on misleading information at worst.

As the gap in both geographic and mental miles between consumer and store bought food has widened, the role of product promotion as a source of product information has grown. Although truthful advertising is more important now than ever, it is not always treated that way. This book exposes both major and minor marketplace deceptions and the damaging consequences of government and public inattention to them.

The book is divided into four parts. The first follows the early years of Florida's orange juice industry. The second focuses on the beginning of federal orange juice regulation: the 1961 hearings into developing standards of identity for orange juice. The hearings show how the FDA, orange juice processors, and comparatively few consumer representatives met in 1961 to shape, through their definition of the standards, consumer knowledge and thus consumption of processed orange juice. The third part traces the development of Florida's orange juice industry post–FDA standardization. It highlights the effects of the processor's power to control the modes of producing not only orange juice but also orange juice knowledge. Part four considers the future of Florida's orange juice industry and what it has to say about trends in agricultural production and processed food consumption in the United States. The concluding chapter returns to the question of the role of a consumer right to know when not knowing is not fatal.

Abbreviations

CREC:	Citrus Research and Education Center
FCOJ:	Frozen Concentrated Orange Juice
FDA:	Food and Drug Administration
FDOC:	Florida Department of Citrus
FFDCA:	Federal Food, Drug, and Cosmetic Act of 1938
NFC OJ:	Not From Concentrate (a.k.a. "pasteurized") orange juice
NRC:	National Research Corporation
Recon:	Reconstituted (from frozen concentrate) orange juice
RTS:	Ready to Serve
SRL:	Subsistence Research Lab
USDA:	United States Department of Agriculture

I
The Growth of Florida's
Orange Juice Industry

I

The Seeds of Florida's Sunshine Tree

Say "orange juice," and Florida comes to mind. The Florida stamp on a carton of orange juice is considered a sign of quality. Florida is where orange juice is supposed to come from.

Florida was also once synonymous with oranges. Not anymore. Now say "Florida" and Disney World, golf courses, and condominiums come to mind. Florida's transformation in the national imagination reflects reality. Orange trees in Florida are relatively few and far between. They no longer line the highways as they used to, sprouting juice stands along the way. Whole groves are being uprooted to make room for the state's tourists and retirees. Come-visit-Florida ads show pictures of surf and golf turf rather than groves and truckloads of oranges.

The simultaneous view of orange juice as the territory of Florida and Florida as mostly grass, concrete, and saltwater is puzzling. How can a state that grows a marginalized number of orange trees supply the rest of the country, continent, and world with top-notch orange juice, as many consumers still believe? It cannot. Although roughly 96 percent of Florida's oranges are processed into juice, the actual number of juice oranges the state grows is declining. Oranges from Brazil, not Florida, supply North America and the world with most of its juice.

The association between orange juice and Florida is not baseless. Florida originally grew most of its oranges for the fresh market. That changed when processed orange juice, with the state's help, became a North American staple. After more than a century of transporting oranges whole to states across the nation, Florida turned almost overnight into the home of the juice orange and orange juice industry. The 96 percent statistic is a relic of this time not too long ago when orange juice and Florida were one and the same. Taking a trip back is essential to shedding outmoded beliefs about the current state of the industry.

Not even Californians, who can claim a few of the earliest citrus processing plants, would argue that Florida is the birthplace of processed orange juice. The industry that grew there was centuries in the making. First the orange had to find its way to North America.

Florida's oranges have traveled far from their birthplace in Southeast Asia. *Citrus sinensis*, the sweet orange upon which Florida's orange juice empire is built, is thought to be a cross between a pummelo, a seedy thick-skinned ancestor of the grapefruit, and a mandarin. The hybrid, which makes it a "convenience" rather than true species, is a native of northeastern India and the adjoining portions of China and Burma.[1] From India the Arabs helped spread the sweet orange to Europe and northern Africa in the fifteenth century. Oranges rolled through

Persia, Syria, and Italy, landing in the south of France. The bitter orange, a close relative, took a slightly different route. It moved from Arabia to Egypt, continuing through northern Africa to Spain and Portugal.[2] As John McPhee noted in regards to the orange's westward journey, in the sixth and seventh centuries Islamic forces conquered the world from India to Spain, growing trees along the way.[3]

The English word "orange" derives from the French "or," meaning gold. The name suits a fruit that, in seventeenth- and eighteenth-century Europe, was a novelty and restricted to royalty—Louis XIV reputedly loved oranges. It is also fitting for a fruit that has for centuries, and especially upon its arrival in Florida, proved to be worth at least its weight in gold.

Southern China was the first to grow the sweet orange commercially. Cultivation in the upper Mediterranean basin began in the mid-fifteenth century. The seeds for Florida's prolific Sunshine Tree were not planted until 1560 when, with golden fruit in hand, Spanish explorers began frequenting the New World paradise. Pomologists, those studied in fruit trees and their cultivation, assume that the original trees that grew in Florida were of the bittersweet type that was common in Seville. Although a long time arriving in Florida, the orange took readily to its new semitropical surroundings. The bitter seeds from Spain began to grow into sweet commercial groves after Spain ceded Florida to the United States in 1821.[4]

Historians of the Florida citrus industry trace its beginnings to the grove that Captain D. D. Dummitt planted on Merritt Island in 1830. Dummitt's grove, believed to be the oldest in Florida, authored the Indian River citrus label that still carries a national reputation as a premium product.[5] The construction of railroads during the 1860s helped push the industry that Dummitt started forward. However, the real stimulus for the orange industry awaited the introduction the following decade of two sweet orange varieties that colonized Florida in the twentieth century.

The first of Florida's two dominant orange varieties arrived in Florida via a circuitous, and among historians disputed, route. The standard account has citrus enthusiast Thomas Rivers bringing the variety from the Azores to his world-renowned nursery in Hertfordshire, England, in 1865. He catalogued it, so the story goes, under the name Excelsior. Entrepreneur Samuel Parsons then took over the variety's journey from Portugal to Florida. A Yankee from Flushing, New York, Parsons established Florida's first orange nursery in Lakeland in the 1870s. He brought the variety from Rivers's nursery in England to Florida. He proceeded to give the imports to E. H. Hart of Federal Point Florida for distribution. Confusion followed. After losing the label, Hart made up a name that suited the trees' agricultural cycle. Consistent with their comparatively late peak date between March and June, he tagged them "Hart's Late." The name changed once in trade to Hart's Tardiff.[6]

According to one version of what transpired next, simultaneously, yet in a distant time zone, a Mr. Calby A. Chapman of California had also been patronizing Rivers's nursery in Hertfordshire. He, like Parsons, was seeking new orange varieties with which to build his citrus business. Between 1870 and 1872 he imported a number of varieties. One supposed navel was not what the label said. Word has it that, acting on the suggestion of his Spanish laborer, Chapman named the mystery variety Valencia Late.[7] Variations on this account abound. The events unfold differently depending on the dramatist's leaning. Under the penmanship of historian of Florida's citrus industry, Robert Hodgson, Chapman did not acquire the late maturing trees from Rivers's nursery. Instead, in 1876 Parsons sent some trees to Chapman, one of which was called "Rivers' Late." In this twist, a Spanish visitor noted the variety's similarity to a late maturing orange that grew in Valencia. He, not Chapman's laborer, renamed the tree Valencia Late. Eventually it became evident that Hart's Tardiff and Valencia Late were the same fruit. So Hodgson explains the name "Valencia" for the variety of orange that almost all in Florida agree, and in the words of Jim Griffiths, managing director of Florida's Citrus Grower Associates, is "far and away the best thing we grow."[8]

From Hodgson's Florida-centered perspective, Chapman's Californian involvement does not jeopardize Florida's claim to being the first American home of what has become known as the Cadillac of oranges. But Hodgson is not content to end the matter of the gold standard of oranges there. He challenges the insinuation made by the name that the variety derives from Spain. He cites Madrid agronomist Gonzalez-Sicilia, who hypothesized that Florida soil developed the variety's identity forming characteristics that the Spanish then imported neatly packed within the orange skin's wrap.[9] Florida is, in Gonzalez-Sicilia's books, not merely Valencia's first *American* home. Florida is *the* home and native land of the orange that is prized today by every juice processor for its deep orange color, distinctive flavor, and high juice content.

The origin of Florida's other distinguished juice orange, the Hamlin, is less controversial. The Hamlin orange entered Florida's citrus industry through the common practice during the nineteenth century of planting trees from seed. The orange's hybrid nature means that planting the seed of one variety will not guarantee an identical offspring. This unpredictability made studied walks among trees a necessity in the early days. On the one hand the random genes contained within each seed could turn up trees as valuable to the commercial grower as a weed. On the other hand they could combine to produce fame and fortune, as they did for H. E. Hamlin. During one walk in 1879 Hamlin discovered a chance seedling in his Glenwood Florida grove that had favorable qualities. He called the variety Hamlin.[10] The rest is history.

In the 1880s budding rather than planting from seed became the propagation method of choice among commercial growers. The technique promised earlier bearing, less thorny, trees holding uniform fruit. According to the standard procedure that survives today, each tree is conceived in a nursery, where seed for the tree's rootstock is planted in a tray. In three to four months, when the seedling stands about six inches high, it is transplanted into a field or, in the case of the self-explanatory "container" method developed in the 1970s, a pot. Whether in field or pot, the seedling is allowed to rest long enough to grow strong enough for the grafting operation known as "budding." Two months is the norm. When ready, an incision is made in the seedling's stalk where budwood, now commonly ordered from Florida's Department of Plant Industry, from the sweet orange variety of choice, is inserted. The budded tree takes an average of eight months, for a total of a little more than a year from the start date, before it is ready to sell to the grower.[11] The tree the grower receives from the nursery does not begin bearing fruit immediately. At least five or six years are needed before the tree begins to produce commercially.

Budding transformed citrus growing in Florida. Since the introduction of the procedure, the sweet orange has had less opportunity to express the genetic diversity contained beneath its peel. Each commercial orange tree bears the lesion where carefully screened and selected budwood of a particular variety was surgically inserted into the rootstock of equally scrutinized pedigree. The resulting uniformity allows for larger groves and less grower interaction with single trees. Whereas a hundred-acre grove would have been considered large in the 1880s, now groves spanning thousands of acres are not unusual.

Roland Dilley, the owner and founder of one of Florida's largest citrus nurseries, describes the situation. "There's not very many horticulturists around. They're money counters, some of the sharpest kids I've ever seen with a computer. . . . But to grow you need to make tracks in what you're doing, without footprints [you're] in trouble. You better be out there looking."[12] The new super-sized groves have made the types of searches that growers such as Hamlin daily performed not only infeasible but also unnecessary. Now that every tree is for the most part the same—an errant tree will pop up infrequently as nature's revenge—there is little reason to walk among them.

The end of Hamlin's routine marks the loss of a major source of sweet orange diversity. Unlike the past, growers are no longer introducing new varieties into *Citrus sinensis* lexicography. What this will mean for the hybrid's future is unclear. Michael Kesinger, chief of Florida's Bureau of Citrus Budwood Registration, a branch of the Florida Department of Agriculture's Division of Plant Industry, is hopeful that "new improvements . . . are going to be introduced by citrus breeders." He is in a position to know. The Bureau tracks new *Citrus sinensis* introductions as part of a program to prevent the spread of disease, a hazard of the grafting procedure.[13] Yet decades of citrus research and generations of breeders have yet to provide a match for Hamlin's keen eye. The Hamlin tree still stands uncontested as the principal early maturing sweet orange variety. In the 2002–2003 season it even surpassed Valencia in popularity. The pale-colored but highly productive Hamlin held an impressive 44.2 percent share of the sweet oranges propagated in Florida, while the Valencia held 34.8 percent. Together the two nineteenth-century introductions comprised almost 80 percent of Florida's sweet orange crop.[14] Breeders will have to work quickly to make a dent in this figure anytime in the near future.

II

The Twentieth-Century Squeeze

Although the raw materials were available at the end of the nineteenth century to supply a booming late twentieth-century orange juice industry, they were not immediately squeezed to their potential. For the first half of the twentieth century, Florida's citrus industry was geared toward the production of fresh fruit. The Valencia and Hamlin, ideal for making juice, were less remarkable as fresh fruit. Many of their attributes, such as few seeds, though assets to the modern processor, were a liability to turn-of-the-twentieth-century growers. Those who had not yet converted to grafting preferred varieties that had seeds with which to grow more trees.

The conclusions of a committee of citrus growers that met in 1915 reveal how little attention the Hamlin initially received. The committee suggested that growers restrict their plantings of sweet oranges to Parson Brown, Homosassa, Pineapple, and Valencia.[1] Although there was no late-season equal to Valencia, evidently Parson Brown, a seedy early maturing variety, was preferred to Hamlin well into the twentieth century.

The recommendations of the committee ring discordantly in the ears of the twenty-first-century juice processor. Processors now know how poorly Parson Browns perform next to Hamlins. Tropicana has nothing positive to say about the once heavily planted variety. The company used the variety in its blend until 2000, when extensive testing demonstrated its negative effects on juice quality. It has since pulled the Parson Brown out of its "Pure Premium" line. As of 2004 it was looking forward to the expiry date of its contract with a high-volume supplier of the variety.[2]

Not all the committee's words were inimical to the growth of a thriving juice industry. Its suggestion to pare down plantings to four varieties reflected a growing belief that the diversity of existing groves was inefficient. A state nursery inspector echoed this concern at the 1922 meeting of the Florida State Horticultural Society. The inspector praised the USDA pomologist who spoke before him for having "shown clearly the handicap under which Florida, with many varieties of oranges, labored when competing with California, which has but two." The pomologist, the inspector reminded his audience, had emphasized "the advantages of the smallest possible number of varieties to cover the season."[3]

Florida growers eventually listened. The Parson Brown, Homosassa, Pineapple, and Valencia lineup became a common rotation. The first is an "early" orange, the second two "early-mids," and the last a "late" bloomer. Together the foursome promised fruit every month of the season while limiting the number of varieties to no more than two at any one time. This was the closest Florida could come to California's dynamic duo, the navel and Valencia.

For the juice processor Florida's inability to mimic California's cycle exactly was propitious. While only one of the two common varieties grown in California was suited to juice processing, each in Florida's foursome boasted a relatively good juicing profile. Florida was primed to become the world's leading juice orange producer.

The redesign of Florida's groves according to the 1915 committee's advice pushed the state's orange industry, already running in overdrive, toward juice dependency. By the turn of the twentieth century Florida's citrus groves had grown out of control. The growers knew it. Beginning in the twilight decades of the 1800s, each year they rode on horseback from all corners of the state to whichever town was hosting the annual Florida Orange Growers Convention. The Convention provided a rare opportunity for citrus growers to unite to discuss the condition of their industry.

In 1909 one issue was on everybody's mind. Production had reached six million boxes, up a million from the pre-1895 five-million-box-per-year average. Mulling over the numbers, the growers considered their options. To prevent oversupply from devaluing their fruit, they could intervene to control either supply or demand. They could ally themselves to curtail production on a statewide scale, or they could join to create marketing procedures that would keep demand, and therefore prices, high. They chose the latter. After a series of committee meetings that followed the June 1, 1909, Convention, a new entity was born. On July 22, 1909, the committee's chairman welcomed into existence the Florida Citrus Exchange as "a living acting being. It is here for all time to stay, for the benefit of the growers of Florida, and will continue under their control," he said optimistically.[4] The Exchange held hope that the state's citrus growers could stay in business without having to cut down their literally overbearing trees.

During the first half of the twentieth century the fresh fruit, rather than processed juice, industry controlled the mar-

keting of Florida's oranges. In 1915 the first U.S. citrus processing plant opened in Haines City, Florida. This plant, and the
ones that followed, served the needs of the fresh fruit industry
by providing an outlet for "eliminations," the term still used
for fresh citrus that fails to make the grade. One of today's
largest processors of pasteurized juice, Florida's Natural, is a
cooperative that originated as one such outlet for second-rate
fruit. Six to seven fresh fruit packinghouses organized the cooperative in 1933 to deal with the fruit that they could not use.[5]

The canned juice that the early processors packed left
much to be desired. The juice could not compete with the taste
of freshly squeezed. Canneries thus never held more than a
tiny share of Florida's citrus industry. In 1930 Americans drank
0.01 pounds per capita of canned orange juice, the only widely
available processed orange juice. The same year Americans consumed 18.9 pounds per capita of fresh oranges.[6]

As the Florida Citrus Exchange grew more ambitious,
the prospects for processed citrus grew more auspicious. James
Hopkins, historian of the Florida Citrus Exchange, marks
1916–17 as the beginning of the Exchange's involvement in serious demand side manipulation. Its approach to overproduction heralded an industry built on gimmicks to encourage
consumers to eat more citrus. In 1916 the Exchange began
earnestly toying with "the first of what would prove to be a
long list of premium gadgets aimed at inducing the American
housewife to purchase more citrus fruit."[7] The device in question was a fruit juice extractor. At first the extractor did not go
over well. Skepticism among some Exchange members as to
both the ability to manufacture a low-cost extractor, and the
Exchange's role as merchandiser, put the idea on hold. Then
the shadow of war, combined with the historic freeze that
killed off many trees in 1917, turned the Exchange's attention

from building consumption to increasing production. In response to a USDA message announcing the wartime duty of Florida citizens to increase their efforts to produce more staple food products, the Exchange's board of directors passed a resolution to grow more citrus. *Citrus sinensis*, as of 1915 planted in the United States at a ratio of five to every one lemon, acquired special status.[8] It officially became a staple, its production patriotic.

Producing an adequate supply of oranges was never more than a short-lived problem for Florida during the first half of the twentieth century. Underproduction was almost always a result of the freezes that periodically descend on Florida. During its early years Florida's citrus industry experienced many, especially in 1894–95, 1899, 1917, and 1935. Although severe, none seriously altered the shape of the industry. Prior to the freezes of the 1960s and '80s, when the nation's thirst for orange juice necessitated a backup source of oranges, growers recovered without outside intervention.

Overproduction was the chronic condition that required the Florida grower's constant attention. It was not long before the idea of a juice extractor for every kitchen was reconsidered. By September 17, 1924, dissension within the Exchange over logistics had disappeared. The board unanimously agreed to initiate a program to distribute juice extractors on a premium basis throughout the nation. At the close of the 1926–27 season extractors etched with the Exchange's "SealdSweet" signature, Florida's answer to the Californian "Sunkist" seal, had made their way into the homes of 16,324 consumers. A report on the program painted a bright future: "It can be safely said that the sun always shines on the SealdSweet extractor. . . . In effect, the extractor is functioning as an advance agent of the SealdSweet brand. Each extractor bears the stamp SealdSweet and in

this is a medium of introduction and a continuous silent sales-
man for our brand."[9]

The report did not forecast the clouds on the Florida Citrus Exchange's horizon. By urging consumers to drink their
oranges, the introduction of the extractor into American homes
helped stimulate the demand for the ready-made juice that eventually made the Exchange, and its kitchen extractor, obsolete.

While the Exchange was working to boost demand for fresh
citrus, another grower-based organization was gaining promi-
nence. To help themselves deal with the economic hardships of
the depression, Florida growers moved to create an organization that became an official government agency in 1935 and that
remains influential today as the Florida Department of Citrus (FDOC). The FDOC, in step with the demands of World
War II, shifted the priorities of Florida's citrus industry. Providing the civilian population with fresh fruit no longer topped
the list. Figuring out how to supply highly mobile troops, operating throughout the globe, with agreeable juice became the
focus of attention.[10]

The Subsistence Research Lab (SRL), the branch of the
Quartermaster Corps responsible for developing foods for the
soldiers during World War II, wanted to include a good tasting
citrus juice as part of its ration formulation because the group
was committed to filling the soldiers' nutritional needs through
diet instead of vitamin supplements. As one colonel noted in
August 1942, the "use of vitamin capsules or tablets . . . was declared to be impracticable." A year later an army captain reinforced the army's reluctance to use vitamin supplements: "For
various reasons we shun vitamin pills for routine administration. We want our nutrients in our foods."[11] The army's resistance to pills created a tall order: an easily transportable and

tasty juice that would function as the primary vehicle for the delivery of a key nutrient, vitamin C.

Soldiers had already been receiving citrus in powdered form. By 1943 lemon crystals were a critical component of the famous K ration, which was the product of a 1941 request to the SRL for three small food units that could fit into the pockets of a parachutist's pants. The result was the "Parachute and Mobile Troop Ration." In the fall of the same year the ration's appeal had become so widespread that representatives of the Quartermaster General and the Chief of Infantry gave it the more generic name "K ration." The SRL's deliberation over each item in the ration was evident in the ration's intricate composition. The ration packed not only synthetic lemon powder for vitamin C, but also eggs and cheese for vitamin A, and biscuits containing whole wheat and soy flour, wheat germ, dried milk, and whole eggs for the vitamin B complex. It also contained a meat component for protein, and an assortment of other products, each of which served a discrete purpose. All told, the ration hovered around three thousand carefully distributed calories—18 percent from protein, 35 percent from carbohydrates, and 47 percent from fat—concentrated in a 42.5 ounce package.[12]

The K ration's nutritional balance led subsistence historian Walter Porges to praise it for carrying "a better assortment [of vitamins] than many civilian diets." He described the meal for the soldier in the field as a "dietetic microcosm" in which each of the components complemented each other, from their interlocking positions within the box to their nutrition and variety of flavor. There was only one sticking point in the smooth operation of the nutrition machine: palatability. Reports from the front lines indicated that the soldiers particularly "detested" the lemon crystals.[13]

Franklin Dove, pioneer of the science of food acceptability and eventual chief of the food acceptability branch of the SRL, a 1944 add-on to the SRL, documented the waste that resulted from unappetizing army food. He wrote about how "certain rations paved the roads to the front lines and filled the gutters along the detours," and how they began to accumulate "in the storage dumps in the various theatres of war." World War II historians Charles Romanus and William Ross theorized that no amount of counseling could convince soldiers to down foods they disliked, because American soldiers came from a place where "availability alone is not a compelling argument for eating."[14] Regardless of why the soldiers rejected their rations, the SRL knew that citrus was useless to the soldiers unless pleasant tasting. So began the quest to make a palatable citrus juice.

The year 1948 was like no other in the history of Florida citrus. After almost a decade of research a group of scientists honed a process for making a prepared orange juice that was better than the canned juice that had been around since the turn of the century. To many, the boiling that was performed for canned juice's preservation evaporated the flavor. Florida citrus history enthusiast Thomas Mack said that the final product was christened "battery acid."[15]

The events that culminated in the 1948 discovery of a marketable frozen concentrate were under way in the early 1940s. With the support of the federal government and recently established Florida Department of Citrus, a group of scientists assembled in 1942 to develop a citrus product that would meet the army's needs. The group's home base was the USDA's Citrus Products Station in Winter Haven, Florida. Three members of the group were ultimately enshrined in the Citrus Hall of Fame for their work with processed orange juice. Louis G.

MacDowell, the research director of the Florida Citrus Commission (part of the FDOC), led the team and did not want its efforts to end in a mere journal publication. From the start he solicited the backing of industry. Edwin L. Moore had a strong chemistry background and, as an octogenarian, was still frequenting his FDOC office as of 2004. Cedric D. Atkins was an engineer by training.[16]

Moore, who grew up in Massachusetts, was the most unlikely to make a career in citrus. He traces his interest in citrus to his college years, when he was far from the Sunshine State. While working part-time at an A&P grocery store he observed deliveries of orange crates with rotting fruit inside. The oranges from Florida tended to have thin skin, making them especially susceptible to decay. Moore grew up in a family of modest means and for whom oranges were considered a Christmas or Thanksgiving treat. He hated to see them go to waste and remembered thinking that "there must be a better way." A slight stutter always made him partial to research, and after college graduation he took civil service exams in chemistry and another in USDA processed foods inspection. He qualified for a Florida Citrus Commission–financed position in Winter Haven.[17]

When Moore arrived in Winter Haven, World War II had begun and the enlistees were in desperate need of a good source of vitamin C. After examining powdered citrus, the MacDowell, Moore, and Atkins team began testing concentrate. They were not the first to study the potential of this technique; over the years citrus processors and scientists had tinkered with the dairy industry's technology for making condensed milk. The early attempts to adapt dairy methods to orange juice exposed the latter's delicate nature. The golden liquid was heat-intolerant, as high temperatures burned off its shine and produced a viscous and brownish mixture that lacked fresh flavor. The outcome was essentially orange—technically, brown—syrup that was then frozen.

In 1943 the trio began to look more critically at the traditional "hot-pack" process for concentrating orange juice. They broke the process into the component steps. The standard approach was to evaporate the water from fresh orange juice in a vacuum at a temperature below 80 degrees Fahrenheit. The resulting high-density solution was then canned and frozen. Consistently finding the reconstituted solution sickly without doctoring, they kept experimenting. Realizing that fresh, full-strength orange juice tasted just as good after freezing as it did before, they decided to try adding fresh orange juice to the concentrate coming out of the evaporator. The "cut-back," as it became called—adding the fresh-squeezed juice decreased and therefore cut back the concentration of the evaporated juice—was a success. It both restored the fresh orange flavor and standardized the percentage of orange sugar solids in each container.[18] Cut-back made each can of frozen orange juice concentrate an exact replica of the previous one. The process was patented in 1948, and the *Tampa Tribune* excitedly pronounced that "FCOJ" was "bound to put millions of dollars eventually into the pockets of citrus growers."[19] The reporter underestimated FCOJ's potential; the product stimulated the orange plantings that would eventually grow into a crop with an on-tree value reaching close to one billion dollars at the close of the century.[20]

MacDowell's early networking with industry must partly explain the fluid transfer of the government-owned cut-back procedure to the private sector. Preparations for the handover were already being made in the mid-1940s: the National Research Corporation incorporated Florida Foods in 1945 to complete construction of a plant in Plymouth, Florida, for the commercial production of frozen concentrate. Florida Foods then entered into a contract with Vacuum Foods Corporation whereby it ceded the right to use the cut-back process to the company. At the end of the decade Vacuum Foods was renamed Minute Maid.[21]

In addition to Minute Maid, Tropicana also benefited from the team's expertise. Atkins once described the location for the processing equipment he helped design for the pasteurized juice company: "We started on a dirt floor in an old fish factory."[22]

In the 1947–48 season the Plymouth plant was one of three churning out a total of 1.7 million gallons of concentrate. The next year seven more plants were in operation. Together, the ten plants packed an estimated ten million gallons that season. Unfortunately, all this occurred too late for the war effort. According to an army report, an estimated 24 percent of ration vitamin C content derived from citrus products.[23] Had FCOJ arrived earlier, it most certainly would have made citrus a more major contributor.

Too late for the war, FCOJ did arrive just in time to address a crisis within the citrus industry itself—which, at that point, according to Thomas Mack, "was just before dying." Sharon Garret, who with her husband owns a three-hundred-acre orange grove in Florida, has been very involved in the local grower community and asserts that pre-concentrate growers "were starving until concentrate came along."[24]

The Florida Citrus Exchange's 1938–39 annual report supports the view of an industry in trouble before FCOJ's arrival. As of May 1939 Florida had produced ninety-four million boxes of citrus, a 25 percent increase over the prior season. Historian James Hopkins concluded from his review of the Exchange's report that year: "High volume output of citrus had caused citrus consumer prices to dip to a far greater degree than the corresponding dip in consumer income. Thus, while the consumer had money to pay for citrus at reasonable prices, the Florida citrus industry had persisted in so glutting the markets that prices were not held in line with the consumer's ability to pay." The Exchange accordingly declared the 1938–39 season "a record of distress." In its report it warned: "Only those will survive and continue uninterrupted grower service who best are able to get the market for fruit handled at the least cost for the service performed." The report advised that the grower "must obtain as big a percentage of the gross delivered return for his crop as is possible if he is even to come close to recovering production costs."[25]

The Florida Citrus Exchange had failed to save growers from a surplus of oranges. Despite the Exchange's efforts to build demand for Florida citrus, orange growers were still struggling with the excess that was slowly starving the hands that fed it. Growers looked to FCOJ to do what the Exchange had not. Fourteen years after being patented, the juice, still in its youth, appeared capable of reviving an ailing industry. At the 1962 gathering of the Florida State Horticultural Society, MacDowell, Moore, and Atkins offered promising statistics. They quoted a USDA report that said oranges going to frozen concentrated orange juice returned 49 percent of the consumer's purchase price to the farmer. This percentage compared to 30 percent for fresh oranges, and 37 percent for oranges that went to canned juice. These numbers led to the team's hope "that the increasing use of frozen concentrated orange juice, either in its present form or as a higher density product, will be able to continue to absorb our constantly increasing production at favorable returns to the grower." FCOJ became dubbed "The Cinderella Product." Florida provided an ideal environment for FCOJ to work its magic. The state has always had a reputation for growing the thin-skinned, juicy oranges that are ideal for processing. With the best juicing varieties, Valencia and Hamlin, already a part of the grower's vocabulary, Florida was ready to churn out orange juice.[26]

FCOJ appeared to be the perfect fit for not only the grower but also the consumer. In 1952 the American Can Company advertised that frozen orange juice had, from 1951 to 1952, saved housewives the equivalent of fourteen thousand years of drudgery.[27] If FCOJ seemed to save the day for the grower, it was promoted as saving many more for the consumer.

In the beginning the Cinderella analogy seemed apt. FCOJ sponged up the Florida oranges flooding the markets of the late 1930s and early 1940s. It brought orange growing back into a means of making a decent living. But the remedy of FCOJ for Florida's orange overload was temporary. Everybody in the industry agrees on FCOJ's ultimate effect. Moore puts it plainly: "Growers started to plant trees and plant trees and plant trees."[28]

The huge increase in orange production that took place in Florida between the 1943–44 and 1951–52 seasons is proof that Moore was not exaggerating. Assuming growers reacted to the buzz in 1943 that a marketable frozen concentrate was in the making, their juice-inspired plantings would have taken many years to bear fruit. The orange tree's long gestation period of approximately seven years means that trees planted in 1943 would not make their mark on production until the early 1950s. In the 1943–44 season Florida produced 46.2 million boxes of oranges. At the end of the 1951–52 season the number had multiplied to 78.6 million boxes. The difference in number of boxes produced over this eight-year period can be attrib-

concentrated orange juice grew the most, from the single strength equivalent of 4.74 pounds to just over 16 pounds.[2]

FCOJ inventor Edwin Moore believes that this steep climb in processed orange juice consumption occurred because advertising agencies enlisted the talents of the famous crooner Bing Crosby for product promotion. In the early 1930s radio advertisers experimented with crooning in an attempt to create intimate and entertaining commercials, and it soon became the rage in radio advertising. In the mid-1940s psychologist Ernest Dichter concluded from his radio research that the public desired more commercials delivered as "Bing Crosby does it."[3] FCOJ processors responded, and Crosby became their man. The apotheosis of crooners, he produced a loyal following of frozen orange juice concentrate consumers in the 1950s.

According to Moore, Minute Maid did not have enough money to advertise, so Crosby agreed to do commercials in exchange for cheap stock options in Minute Maid. In 1948 Crosby sang: "There's no doubt about it"—Minute Maid, his ditty went, is "The best there is," and he extolled the juice for its "fresh squeezed" taste. The deal Minute Maid and Crosby sealed was mutually beneficial. "Bing, he got rich on that and of course Minute Maid took off too," Moore said. He has no doubt that marketing was a major factor in FCOJ's almost instant transformation into a beverage favorite.[4] Advertising historian Roland Marchand theorizes that crooners became such popular marketing tools during the 1940s and 1950s because advertisers believed that a crooner's therapeutic tone suited a society struggling with the increasing complexity and impersonality of modern life. They solicited the likes of Crosby in response to signs that consumers wanted to be "amused rather than instructed."[5]

Crosby's songs in praise of Minute Maid paved the way for a new commercial age in the United States. Americans are now exposed to an average of as many as four hundred product advertisements daily on TV, radio, newspapers, and magazines. Communication studies expert Matthew McAllister notes that the United States has become "ad burdened," as it constitutes less than 10 percent of the world's population but accounts for 57 percent of the world's advertising spending.[6] As a result of the glut of advertisements product information is routinely sacrificed for entertainment. In the case of orange juice, TV ads show oranges bopping to music, juice cartons portray oranges air-brushed to perfection, and tantalizing images and sweet sound bites proliferate while not even the fine print on orange juice labels fully discloses what's inside.

By 1960 the Food and Drug Administration (FDA) had amassed a number of complaints and recognized the confusion that orange juice packaging and promotion was generating. The agency objected to processors adulterating orange juice with water and sugar, and it was concerned about the misrepresentation of processed orange juice as "fresh." In 1956 the FDA, along with Kraft Foods Company and the National Association of Frozen Food Packers, proposed to set standards for orange juice products. A flood of objections to the proposed standards prompted the FDA to hold hearings to determine how to define orange juice standards of identity. What took place during those hearings constitutes part two of this history of Florida's orange juice industry.

II
Developing Orange Juice
Standards of Identity

IV

Introducing the FDA
Standard of Identity

*Unfortunately the food products about which consumers often
know the least are those which have been standardized.*
—*Faith Fenton, Cornell University professor of
food and nutrition, 1961 FDA hearings*

The 1938 Federal Food, Drug, and Cosmetic Act introduced the "standard of identity" into the FDA's vocabulary. The Act provided the FDA with the authority to set a standard of identity for a food "whenever in the judgment of the Secretary such action will promote honest and fair dealing in the interest of consumers." Food processors had been marketing ice cream with varying levels of cream, maple flavored syrup with varying levels of maple sugar, and jams and jellies with varying levels of fruit. With a standard of identity the FDA could define the processes and ingredients that it deemed acceptable in a food's manufacture. By establishing a standard of identity for "maple flavored syrup," for instance, the FDA effectively prevented manufacturers of almost 100 percent corn syrup from selling their product as maple syrup. Ultimately the standards were supposed to ensure that consumers received good food value for their dollar.[1]

Tomato products were the first to be standardized. Jams and jellies followed. By 1957 the FDA had standards of identity for most of the foods found in a well-stocked pantry: chocolate, flour, cereals, macaroni and bakery products, milk and cream, cheese, butter, mayonnaise, canned fruits and fruit juices, fruit preserves and jellies, canned tuna, eggs and egg products, margarine, and canned vegetables. The standards of identity for jams and jellies, which were based on cookbooks and family recipes dating back two hundred years, served as the blueprint for other processed products. Evidence from cookbooks and any other relevant source was typically advanced through lengthy hearings. The hearings for peanut butter, which lasted ten years and produced eight thousand pages of material, finally signaled a need for reform. In 1953 the procedures were amended so as to waive a hearing when there was no controversy.[2]

When there *was* a dispute the fact-finding process sometimes became ridiculously, even comically, detailed. During hearings into developing a standard of identity for enriched white bread, for instance, bread manufacturers petitioned for bread softeners to be included as acceptable ingredients in the standardized recipe. The FDA objected on the grounds that the additives deceived the consumer into believing the bread was fresh. Both sides solicited the social sciences to determine the kinds of judgments that consumers made about freshness

when squeezing a loaf of bread.[3] Judging from how readily and
reliably a loaf of Wonder Bread yields to touch, the bread makers were the winners.

Until 1990 the various standardized products that evolved
from the 1938 act had a common characteristic: their labels had
to list only optional, not mandatory, ingredients. According to
legal scholars Richard Merrill and Earl Collier Jr., the FDA did
not require ingredients lists on the labels because "the under-
lying premise of recipe standards is that consumers—even if
provided with adequate information—are not equipped to
make value distinctions among products similar in appearance and use, and are therefore unable to protect themselves
against deception."[4] The Nutrition Labeling and Education
Act, passed in 1990, changed the rules, even for standardized
foods. Before the Act the only way consumers could know the
ingredients of a standardized product was to look up the FDA's
recipe in the Federal Code. Now even the ingredients of standardized foods must appear on the food label.

The orange juice standards of identity, which were formed
when the American pantry was changing to include more refrigerated and frozen foods, were among the last to be mod-
eled on the recipe. Merrill and Collier observe that in the early
1970s the FDA began developing "common or usual names,"
instead of codifying rigid recipes for foods. They note that the
newer approach, rather than forcing manufacturers to conform to standard recipes, requires them to make the percentage of a food's essential elements part of the food's name. The
scholars conclude that the evolution of the "'common or usual
name' concept," in "providing information without attempting to control food composition," represents "a sharp departure from the plenary regulation of recipe food standards."[5]

Despite their codification during a transitional period, the

orange juice standards of identity stand as a permanent reminder of the pitfalls of regulation that is built on the negation
of consumer information. As the following chapters show,
even with passage of the Nutrition Labeling and Education Act
a Federal Code search is sometimes still needed to fully know
the ingredients that do and do not go into a standardized food.

Born in the early '60s, the orange juice standards of identity
came into being when food growing and processing technology were progressing rapidly. In an article for the USDA's 1960
Yearbook of Agriculture, Earl Butz, who would soon become
Secretary of Agriculture, spoke of the critical changes under
way: "The decade of the 1960's opens with the march of agricultural science in full stride. Agriculture is changing from a
way of living to a way of making a living. It is changing from a
business of arts and crafts to a business undergirded with large
amounts of science and technology."[6]

The reorganization of food growing into a moneymak-
ing endeavor was part of what Butz described as a "third great
revolution" in agriculture. Although he traced the revolution
to the 1940s, he watched it advancing "at an accelerated pace
during the past few years." Food processing was keeping pace.
He remarked: "The technological revolution in the processing
and distribution of food and fiber has been perhaps even more
spectacular than the technological revolution on the farm."
His consistent depiction of changing methods of food production in militaristic terms was fitting. His talk of a "revolution,"
and his reference to the "march" of agricultural science, were
apposite a dynamic food sector fired by former warfare such as
nitrogen and DDT. The former was turned from gunpowder
into fertilizer; the latter enlisted in the new assault against agricultural pests.

By the 1960s the food industry employed twenty-five

thousand food scientists and technicians, a sizable portion of the eleven million involved in the processing of farm products.[7] Plant breeders in the business of engineering crops for processing joined the new brand of food scientist to transform the American food supply into one that was dominated by packaged food. The textbook definition of processed foods captures the many segments of society that food processing engages. They are "raw animal, vegetable or marine materials [that have been] transformed into intermediate ... edible products through the application of labor, machinery, energy, and scientific knowledge."[8]

By the 1960s, 75 percent of purchased food had undergone some form of processing. The fivefold increase between 1960 and 1976 in frozen potato consumption provides some indication of how pervasive processed foods were becoming. Data for fruit consumption is equally telling. The amount of processed fruit that Americans consumed rose from roughly 16 percent of total fruit consumption in 1919 to around 56 percent in 1998.

The upward trend in processed food consumption in the United States has been accompanied by a growth in the number of food additives. In 1956 the FDA estimated that a total of 704 food additives were in use in the United States. The Food Protection Committee of the Food and Nutrition Board classified 301 of these additives as flavoring agents. The FDA currently estimates that over 3,000 food additives are in use, more than four times the 1956 estimate. Approximately 1,500 of these additives are flavoring agents, more than five times the number the Food Protection Committee reported in 1956. Already by 1978, 80 percent of packaged foods contained flavor additives.[9] All this food-related innovation generated need for additional regulation.

The development of the orange juice standards of identity against the backdrop of a food supply in flux endows them with special significance. The FDA's struggle, played out during six months of hearings, to at once justify and at the same time "fix" orange juice standards, documents how the federal government entered the fray. In providing rare footage of government in action during the high point of a food revolution, the hearings expose the discrepancy between lofty legislative aims and their real-world application.

V

Capturing the Interest of the Orange Juice Consumer

Picture a room full of high-powered lawyers questioning government regulators and industry and consumer representatives in an attempt to determine what orange juice is. Although this sounds like a Monty Python skit, it was actually the backdrop for FDA hearings into the "Matter of Orange Juice and Orange Juice Products; Definitions and Standards of Identity." The hearings, which ran through the first half of 1961 and produced over three thousand pages of transcribed material, attest to the weight given to the search for orange juice's identity.

The FDA convened the hearings after receiving more than sixty letters of objection, mostly from orange juice industry representatives, to notices of proposed rule-making regarding the creation of orange juice standards of identity. The agency's authority to define standards of identity derives from what is now s.341 of the Federal Food, Drug, and Cosmetic Act. This section opens with the imperative "Whenever in the judgment of the Secretary such action will promote honesty and fair dealing in the interest of consumers, he shall promulgate regulations fixing and establishing for any food, under its common or usual name so far as practicable, a reasonable definition and standard of identity." The Secretary's obligation to set standards of identity whenever they "will promote honesty and fair dealing in the interest of consumers" triggered debates over the meaning of "honesty," "fair dealing," and the "interest of consumers" during the orange juice hearings.

In the model bureaucratic world, monolithic institutions implement the black and white letter of the law. In our world, conflicted and conflicting individuals meet to decipher words with more than one meaning. In this chapter of federal orange juice regulation, the FDA's chief lawyer, Bruce Brennan, shouldered much of the responsibility of putting s.341 to the practice of protecting the orange juice consumer.

The twelve orange juice identities that are detailed in the Federal Code are a tribute to Brennan's perseverance. The family portrait portrays prolific orange juice innovation. First in line is "orange juice." Following are eleven processed variations on the formative juice. In order of appearance in the Federal Code they are: "Frozen orange juice," "Pasteurized orange juice," "Canned orange juice," "Orange juice from concentrate," "Frozen concentrated orange juice," "Reduced acid frozen concentrated orange juice," "Canned concentrated orange juice," "Orange juice for manufacturing," "Orange juice with preservative," "Concentrated orange juice for manufacturing," and "Concentrated orange juice with preservative."[1]

The hearings concentrated on five: "Orange juice," "Pasteurized orange juice," "Canned orange juice," "Orange juice from concentrate," and "Frozen concentrated orange juice." The application of s.341 "in the interest of consumers" to the manufacture and marketing of these most popular forms of orange juice became one of the most critical and difficult issues to resolve.

Excluding FDA personnel, only a handful of witnesses appeared at the six-month-long proceedings on behalf of the consumer, for whom the standards were being written, and therefore the task of interpreting the "interest of consumers" was complicated. Nevertheless, the testimony of these few consumer witnesses, combined with the FDA's s.341 mandate to protect the consumer's interest, prevented the hearings from becoming an exclusive forum for industry opinion. The advocates for the consumer's case consisted mostly of FDA officials and members of various consumer organizations. The Federation of the Women's Club and the American Home Economics Association anchored the side presenting critical consumer issues.

The testimony of Faith Fenton, then professor of food and nutrition at Cornell University, was especially influential in addressing the interests of orange juice consumers. Her experience—ranging from World War II research into the effects of freezing, dehydration, and irradiation on the palatability and nutritional value of food, to involvement in the food industry's Institute of Food Technologists—gave her unique insight into the history, makeup, and manufacture of processed foods. As a member of the American Home Economics Association she wore another hat as consumer nutrition educator and new industry product promoter.

The labeling of standardized foods was at the top of Fenton's agenda. She wanted to see more-detailed labels. As a nutrition educator she was concerned about marketers convincing consumers which foods they should buy. According to Fenton, labels should contain the information necessary for the consumer to make an "intelligent choice" among "orange juice and orange juice products." The opinion was consistent with the philosophy of FDA commissioner George Larrick, who, Fenton reminded those at the hearings, had pronounced: "The consumer has a right to know what is in his food."[2]

The language of rights, which is derived from the Constitution, a text as authoritative as it is unread by the public, is liable to pass as highfalutin words apposite a select few. But Fenton saved Larrick's decree from the irrelevance of theory by situating it in the contemporary context that gave it concrete meaning. She observed: "[The] consumer is becoming more and more dependent on food that has been partially or fully prepared for use." She continued: "The USDA estimates that in the next 15–20 years the quantity of processed foods offered to the consumer will increase half again as much as will unprocessed foods."[3] Evidence that the rising use of prepared food was not a fleeting trend added dimension to Larrick's edict. Consumers' growing reliance on the distant producer for food fostered a demand for the right to knowledge concerning the preparation of their food.

The move from home- to industry-manufactured food brought with it not only new rights but also new and shared responsibilities. Larrick's identification of a consumer's "right to know what is in his food" positioned the only rights-providing body, government, in a lead role. Fenton widened the scope of duty by introducing the processor-turned-provider, or "industry." She expected a lot from this performer: "More and more the maintenance of high standards of food [is] coming to rest on industry." At the same time she insisted that "more and more responsibility rests on the consumer to help develop proper standards to protect his food supply."[4]

The three-character play—government, industry, consumer—is still running today. In the six-month production run of the "Matter of Orange Juice and Orange Juice Products; Definitions and Standards of Identity," not one among the trio made an impressive showing. Government was unconvincing in its role as enabler of the consumer's "right to know." The FDA ignored its script when it chose standardization for federal orange juice regulation. Fenton did not let the FDA's oversight go unnoticed. She was critical of the nondescript label that characterized standardized foods. The labels confounded even experts in the field of food and nutrition. Fenton noted that "some home economic people say that since no ingredients are listed on the label, [they] thought no added ingredients [were] present." This hearsay led to her concise summary of the problem with standardization: "Unfortunately the food products about which consumers often know the least are those which have been standardized." She argued that if standardized foods were singled out as not requiring ingredients lists on their labels, their special status should at least have to be made public.[5]

The performance of the other two in the group was equally wanting. The industry far from lived up to Fenton's expectations. Its failure to "maintain high standards of food" was part of the reason that government was acting in the first place. The consumer's showing was also less than glowing. Despite Fenton's assertion that "the desire to know is [a] desirable characteristic of [the] present-day public," the public's thirst for knowledge was barely decipherable in the hearing room.[6] Only one of the seven-odd consumer witnesses who participated in the hearings was a man, and the women who testified were often overpowered by the deeper male voices present.

Although few and far between, words from representatives of the "consumer's interest" such as Fenton did emerge within the flood of industry–FDA banter. Ruth Robinson, who identified herself as housewife and vice president of the Federation of Homemakers, reinforced Fenton's effort to raise the profile of the consumer right to know. She described her organization as a "representative group" of homemakers who incorporated after attending hearings regarding the Food Additives and Chemical Preservatives Act. Those hearings led to the Food Additives Amendment of 1958, which reclassified food additives as prima facie adulterants. Robinson spoke firmly of the housewife's rights: "[We] feel it is our right to have true freedom of choice in the product we buy." Her invocation of the right to free choice in the context of the prepackaged food purchase raised food shopping to the status of constitutionally protected expression. Having planted the right, she stressed the cardinal condition for its germination: "Obviously we cannot have choice unless we have adequate labeling." Her reinforcement of the need for enabling labels reminded FDA regulators distracted by industry opinion of its duty to the consumer.[7]

FDA officials attending the hearings did accept that the housewife had certain rights when it came to food purchases. However, the claim they acknowledged was not the same as the one Fenton, Robinson, and even Larrick maintained. Lowrie Beacham, deputy director of the FDA's Division of Food, recognized a different kind of right when he took the stand to defend the FDA-proposed percentage of orange solids in commercially reconstituted juice: "When [the housewife] goes to the store and buys a product of orange juice concentrate or reconstituted orange juice, where someone else has gone to the trouble to reconstitute the juice for her, [she is] certainly entitled to no fewer orange solids than [she] would get at home if she reconstituted the product herself."[8] While Fenton and

Robinson expressed the food consumer's right in terms of personal liberty, Beacham couched it in the rhetoric of property security. The housewife had title to a uniquely late twentieth-century life necessity. She was entitled not to land to live on, nor even to the fruits of the land to live by, but to food equal to home reconstituted.

The right that Beacham conceived gives new meaning to John Locke's labor theory for private property. In the second of his *Two Treatises of Government*, Locke states the premise upon which his seminal defense of private property is built: "Man has a property in his own Person." He proceeds: "The Labour of his Body, and the Work of his Hands, we may say, are properly his. Whatsoever then he removes out of the State that Nature hath provided, and left it in, he hath mixed his Labour with, and joyned to it something that is his own, and thereby makes it his Property."[9]

Beacham's words may as well be Locke's, adapted to a world where the supermarket is second "Nature." If Beacham's "housewife" substitutes for "everyman," and "supermarket" substitutes for "Nature," Locke's statement now reads: "Whatsoever the housewife removes out of the State that the Supermarket hath provided, and left it in, she hath mixed her Labour with, and joyned to it something that is her own, and thereby makes it her Property."

The resemblance between Beacham's right to reconstituted orange juice and Locke's right to private property is not entirely superficial. Locke's labor theory can be used to justify the orange juice right that Beacham staked out for the consumer. When the housewife Beacham spoke of went to the store, purchased a can of frozen orange juice, took it home, and mixed it with water, she literally mixed "something that is her own" into it. She thereby made the home-reconstituted juice her "Property." There ends the parallels. The rights that Beacham imagined for the housewife in the juice she reconstituted were more expansive than Locke's theory could support: they extended to control over the actions of the processors who mimicked her household moves. The processors' reproductions had to replicate her reconstituted juice exactly, thereby raising Locke's right to untold heights.

The shift in the United States to a diet of mostly—and increasingly heavily—processed foods was not trivial. The substitution of frozen for fresh, and later reconstituted for frozen concentrated orange juice, was a symptom of the substitution of Beacham's orange juice buyer for Locke's laborer, of suburb for homestead. It was also symbolic of a more encompassing dietary overhaul brought about by the migration from rambling farm to prefabricated, reconstituted, white-fenced home. The new North American menu recast basic provisions that were once made and safely contained within the private domain into industrially produced and publicly traded products. This turning of food inside out dispossessed the housewife of her role as domestic food preparer, hence Beacham's fitting, if unwitting, revision of an ancient property regime to provide the 1960s housewife with fresh entitlements in frozen industry prepared foods.

Beacham's description of the housewife's right to reconstituted orange juice as property meshed with the philosophy behind the standard of identity. The housewife's right to a specified number of orange sugar solids was a state-administered right to a state-defined base level of processed juice quality— simply stated, a right to government protection from industry adulteration (or, in the language of property, thievery). The right departed from the self-empowering right to know and freedom to choose that Fenton, Robinson, and Larrick articulated. Still, the recognition that the housewife had any rights was a start to balancing hearings in which industry interests were receiving more than their fair share of the FDA's attention. While Beacham outlined the consumer's entitlements in reconstituted orange juice, Robert Kilburn, director of research and development for the Florida Citrus Canners' Cooperative, alleged a broader right of industry. Kilburn took the stand to tell the FDA why it should not set a minimum Brix—which measures the weight of sugar per volume of solution for a specific temperature—for canned orange juice. In the orange juice industry, Brix refers to the amount of orange sugar, or orange solids, per volume of juice. Brix is still the basis upon which juice oranges are commercially traded. Oranges range from 9 to 14 percent sugar solids.[10] The FDA proposed a 10 percent minimum for canned juice, presumably in hopes of preventing canners from either watering down their juice or enhancing it with synthetic sugar. Although the minimum was lower than the suggested 10.5 percent for pasteurized, and 11.8 percent for frozen concentrated orange juice, Kilburn was still dissatisfied.

He had reason. With the growth in frozen concentrate and pasteurized orange juice technology, consumption of the high-heated, cooked-tasting canned juice with which he worked was falling. Decreasing demand for canned juice translated into reduced bargaining power for its producers. Canneries were left with oranges that had unfavorable profiles, notably low percentages of orange sugar solids, or Brix. As a representative of a sector of the orange juice industry with so little control over the quality of its primary ingredient, Kilburn resisted regulations that would rule out any more material for manufacture. FDA lawyer Bruce Brennan admitted that setting the proposed minimum Brix for canned juice would change industry practices, but Kilburn was more forceful. Such a minimum, he said, would "change our rights."[11] His indictment of the canned juice minimums underscored the industry's inflated idea of its entitlements.

Charles Brokaw, chief of quality control for the Minute Maid Company, fought for industry rights in less explicit terms. He opposed the FDA's creation of orange juice standard of identities on the grounds that "fixed" standards would stay progress. A witness for Edward Williams, counsel for the National Association of Frozen Food Packers and the Florida Canners' Association, he identified recent and significant advances in orange juice processing. He pointed to extraction technology that increased efficiency and essence recovery research that could lead to the substitution of essence for fresh juice in the restoration of flavor to processed juice. He concluded that if the standards currently on the table were set in the early '50s, when the new processing procedures had not been conceived, they would be illegal. The food industry, he cautioned, would be reluctant to invest in research knowing that new standards might arrest the application of research successes.[12] His was a familiar industry refrain: government regulation hinders industry innovation. But Brokaw's focus on the industry's interests in arguing against standardization overlooked the fact that industry was not part of the s.341 equation for determining whether the FDA should engage in standardization.

During an earlier cross-examination, Beacham pursued the flip side of the laissez-fair environment conducive to creativity that Brokaw advocated: the dangers of unmanaged innovation. The opportunity arose when Williams asked Beacham whether there was a difference between partial and almost complete removal of liquid from orange pulp. When extracting juice from oranges processors typically removed the pulp. Williams was trying to discern whether, and in what form, the FDA thought the pulp could be reintroduced. Beacham replied unequivocally that there was a difference. He knew not all pulp was equal. (Allen Morris, formerly in charge of Tropicana's orange and juice procurement, explained in 2004 just how different the types of pulp added back to juice can be. "When you make orange juice you take everything out of the orange so you got the rag and the seed and the pulp and all. And you run it through a drum, a big drum that's a screen and it pushes the juice and it screens the pulp out. Well what's left behind is really a pumice, it's ground seeds and bitter. They put that in water to wash all the juice solids off and that leads to a water solution that is about four percent juice and then they evaporate that." The result is slurry that, he said, "doesn't taste good at all.")[13]

Beacham testified that pulp that had been washed, drained, packed into containers, and frozen was "a little far removed" from pulp that was naturally present in orange juice when first taken out and blended back. He had a simple justification for distinguishing between the forms of pulp that Williams distilled: "The greater the extent the product has been manipulated . . . the greater the possibility for abuse in preparing the finished product."[14] Clearly the downside to unfettered industry advancement was the enhanced opportunity for exploitive behavior.

Beacham was not the only one concerned about industry abuse. Kilburn declared during cross-examination by Brennan that the consumer "accepts on faith the nutritional value of orange juice as a healthful product." Brennan followed up: "We have then a case of the consumer trusting in the good faith of the manufacturer to be putting in ingredients which as you say, are nutritional and which come from oranges?" Kilburn stood by his statement: "Very much." Evidently troubled by the consumer's reliance on taste and the industry's good faith to obtain a quality product, Brennan persisted: "Yet there is a situation which could arise where the processors could violate this trust and yet the consumer would be unknowing, at least as far as appearance and taste is concerned, isn't that right?" Kilburn's response undermined Brokaw's case against regulation: "That is the primary purpose of regulation, particularly the detailed form that it takes in Florida, to eliminate any such possibility." Brennan asked just one more question: "So that then there is reason for regulation in this area so that such a situation does not occur?"[15] Kilburn's "yes" answer ended Brennan's examination. Brokaw's charge that standardization would hold up the juice processing operation lost force under Kilburn's confession that regulation was absolutely necessary. In recognizing the desirability of "detailed" regulation, Kilburn gave the go-ahead to the type of comprehensive FDA regulation that even he had earlier challenged as an undue restraint on industry rights. Federal regulation and orange juice betterment were apparently not, as industry argued at various points during the hearings, mutually exclusive.

An industry representative, Kilburn ironically advanced the FDA's fight to keep the s.341 "interest of consumers" within sight. He indirectly endorsed one of the FDA's basic motives for entering orange juice regulation: what was artistry to industry was potential sophistry to the FDA. The strongest appeal for FDA disengagement—industry's impressive inventiveness— was the very justification for FDA regulation. Kilburn's initial "right to," and Brokaw's "freedom from," arguments backfired. They brought to light the s.341 "interest of consumers" that the industry had been neglecting.

The hearings into the "Matter of Orange Juice and Orange Juice Products; Definitions and Standards of Identity" addressed many issues particular to orange juice. At the same time, the government, producer, and consumer interaction that occurred lent universality to the hearings, as the back and forth between government and industry over the need for consumer protection by federal standards is timeless. Industry complained that regulation diminished its rights; government responded with the consumer's entitlements. Industry warned that regulation would rule out research and development; government responded that without it consumers would be threatened with exploitation at the hands of unlimited industry ingenuity.

While the industry–FDA debate over orange juice standardization revisited an age-old contest over the appropriate reach of government, its intensity reflected a meeting between the two sides that was becoming more heated in the wake of the total makeover of homemaker into consumer.

VI

Regulating Knowledge: The Case
of Pasteurized Orange Juice

By 1955 consumers purchasing orange juice did not
necessarily head straight to the freezer. Pasteurized
orange juice was generating traffic around the refrigerated section of supermarkets. The FDA had one
major problem with the convenient, ready-to-drink juice: it
was not marketed in a way that let consumers know it had been
heat-treated. The common name for the product, "chilled orange juice," lacked any indication that the juice was not, as it
appeared to be, fresh-squeezed, and processors were not re-
ceptive to the FDA's preference of the more revealing name
"pasteurized orange juice." The disagreement was symbolic of
a larger issue in which manufacturers argued the virtues of
maintaining their current processing and marketing practices.
Industry custom to avoid consumer confusion became the processor's mantra against standards that altered the status quo.

The battle over the appropriate name for pasteurized juice
became animated when FDA legal counsel Bruce Brennan in-

troduced the generic "housewife" into the discussion. While cross-examining Horace Campbell, who headed quality control at Sunkist, Brennan asked him whether the "housewife understands that when she squeezes an orange, the juice there from is orange juice?" After Campbell's affirmative response, Brennan continued: "You bring along a product which is heat-treated, kept on the shelf for a longer period than what she squeezes in the kitchen could be kept . . . and you want to call that orange juice too?" "Yes," Campbell said. Brennan was critical: "[You] think it is in keeping with honesty and fair dealing and the interest of consumers to call these two different products both orange juice?" To Brennan freshly squeezed was the original, and any other was not the same thing.

Campbell was equally persistent in pressing for the opposite position: "Yes I think that is what they recognize both products as." He apparently did not believe that processing procedures created unique entities. In his opinion Sunkist's pasteurized orange juice label was therefore in keeping with s.341 of the Federal Food, Drug, and Cosmetic Act. Houston Kier, general manager of California's Glenco Citrus Products, went further in his denunciation of the "pasteurized" label. During his cross-examination by Brennan, Kier testified that calling heat-treated juice "pasteurized orange juice" would create too much confusion. He thought the better solution was to call the heated product "orange juice," and freshly squeezed juice "fresh orange juice."[1]

The idea that freshly squeezed rather than pasteurized juice should be the juice requiring differentiation raised the question of whether a food in its untouched or processed form was the norm, an issue that remained an undercurrent throughout the hearings. Further, because of the recent overall growth in processed food consumption, it became extremely controversial and presented a fundamental point of departure between the FDA and food processor. The argument that pasteurized, not fresh, orange juice should be called "orange juice" assumed that the consumer expected store-bought foods to be processed. FDA officials either disagreed or did not think consumers should have this expectation. Lowrie Beacham, deputy director of the FDA's Division of Food, voiced the general agency opinion. When Brennan asked him whether "fresh orange juice" was the proper name for a product that was freshly squeezed, Beacham said he did not see any "reason for giving [freshly squeezed orange juice] a different name in order to let a new product, a newly developed product, have the traditional name of orange juice."[2]

During an earlier cross-examination of Sidney Brown, a processor appearing on behalf of the National Orange Juice Association, Brennan elaborated the FDA's position. He asked Brown, "Isn't the best way to illustrate a change in the nature of a product to describe it with what brought about that change, the processing?" Brown said he could not come up with an example of a food that had the word "processed" in its name. "You might have heard [of] processed cheese?" Beacham prompted. "[I] don't remember my wife serving me processed cheese though," Brown responded, emphasizing that Mrs. Brown dropped the adjective "processed" in day to day conversation.[3] Brennan did not argue the point.

Brennan was more relentless with Campbell. He refused to accept Campbell's assertion that Sunkist's shelf-stable orange juice should be called orange juice and shifted his cross-examination from heat-treated to frozen concentrated juice. Once again he brought an imaginary housewife into play. He led Campbell into her kitchen, asking: "When [the] housewife takes from the refrigerator a can of concentrated orange juice, reconstitutes it and puts it on the table, what do you think she tells her family this is?" Campbell stepped into the housewife's shoes: "Orange juice." Brennan prodded: "But [she] does know [it is] a different product from [the] one she might squeeze from an orange?" Campbell concurred with a caveat: "I think she knows it is different in the manner she is purchasing it and using it. . . . I don't think she considers it any different from what she gets out of an orange, because she calls it orange juice." According to Campbell, the consumer's descriptor determined product identity; the manufacturing process was incidental.

The speculation-filled exchange between Brennan and Campbell continued:

> BRENNAN: "You mean when she goes to the sink and adds water to a six-ounce can of concentrate, she doesn't think she is getting something different from what she might draw out of an orange?"
>
> CAMPBELL: "I don't think it particularly concerns her one way or the other."
>
> BRENNAN: "Though she might call it orange juice, you think she even thinks in [her] mind that it is . . . the same as what she squeezes out of an orange?"
>
> CAMPBELL ADMITTED: "Hard for me to say what she thinks."

Still, he concluded, "I don't think she is really concerned one way or the other . . . if she was really concerned she would buy fresh oranges."

Brennan persisted. "Then you think buying fresh oranges might be an expression of concern [or] interest in a product which would have characteristics of freshly squeezed orange juice?" Campbell would acknowledge only that this "might" be correct.[4] The question was purely hypothetical; statistics showing growing juice production and declining fresh orange consumption confirmed that consumers preferred to buy their oranges preprocessed into juice.[5] Campbell's comment was thus moot. Maybe the "housewife" would buy more fresh oranges if she were concerned about serving her family processed juice. The question that the proposed standards were designed in part to address was her inability to know whether she *should* be concerned—a lack of knowledge that was the result of the efforts of industry members such as Campbell who wanted to minimize what they told her about their processed alternative to fresh-squeezed orange juice.

The extent of the processors' reluctance to inform consumers about their product became clear during Brennan's cross-examination of Campbell over the subject of heat-treated juice—marketed as orange juice but made from 50 percent reconstituted and 50 percent fresh juice. He wondered whether "orange juice" informed the consumer "precisely" what it was. Campbell answered in the affirmative, adding, "within their understanding." Brennan reiterated this new definition of precision: "So what we are striving for is a definition which precisely informs the consumer within his understanding of what the product is."[6]

If labels only have to provide consumers with information that is within their understanding, producers of pasteurized orange juice did not have to worry. As the FDA's Beacham described it, pasteurized orange juice was a "fabricated product . . . built to specifications . . . treated in such a way as to withstand the vicissitudes of transportation and distribution . . . converted from a strictly perishable product into a semi-perishable product."[7] Given that consumers would not understand the juice they were buying was so "fabricated," precision with respect to product description did not, according to Campbell's logic, require the processor to unmask the truth. Pasteurized and reconstituted orange juice both could rightly be called "orange juice," and processors could be excused from the responsibility of educating the public regarding the complexity of the processed foods fast becoming a dietary mainstay.

In the event that the FDA would not agree to the indiscriminate use of the term "orange juice" for simulated products, processors presented another option. Sidney Brown of Halco Products Inc. offered "chilled" as the most appropriate adjective to identify his company's heated product. He reasoned that "chilled orange juice" was the common name within the trade, the U.S. Department of Agriculture, and even in supermarkets thanks to Florida's "chilled juice" advertisements. He stressed that pasteurized "certainly is not the usual name associated with our product by the consumer."[8]

Brennan later asked Minute Maid's Charles Brokaw whether "pasteurized" was not as good as "chilled" as a referent for the heated juice. Brokaw asserted: "No, the thing that is significant to the consumer is that the product must be refrigerated or chilled for preservation." In his opinion heat treatment was not of "material interest to the consumer." Brennan summarized Brokaw's position: "It's not what has been done but what must be done to the product that is the determining factor?" Brokaw confirmed: "To a certain extent [this] is true."[9] He said that his problem with the word "pasteurized" was that it volunteered information that the consumer did not care to know.

Carroll Brinsfeld Jr., a witness for the FDA appearing on

behalf of the Maryland State Department of Health, related a Tropicana marketing failure that suggested a different reason for processors' protest of "pasteurized." Brinsfeld recalled that sales plummeted instantly when Tropicana once attempted to sell its jarred juice as "sterilized juice." Consequently Tropicana, rather than telling the consumer that its chilled jarred juice was essentially the same as shelf-stable canned juice, called the product "first fresh orange juice with unlimited shelf life."[10] Although Brokaw testified that the *consumer did not care to know* what had been done to her juice, the Tropicana story suggests that the *processor did not want the consumer to know.*

Some within the industry were more readily willing to admit their aversion to names that might put off the consumer. In a letter that Louis MacDowell, then director of research for Florida's Citrus Commission, addressed to the Dairy Service in 1955, he advised how to market juice reconstituted from frozen concentrate: "We do not feel that any useful purpose would be served by designating your product 'reconstituted orange juice.' . . . This wording has a synthetic sound which might very well mislead the buying public." He elaborated: "Frozen concentrated orange juice is well known to the buying public and has wide acceptance in the marketplace. The housewife is firmly aware that she must add water to it in order to dilute it to drinking consistency. The product so obtained is orange juice by all known physical, chemical, bacteriological, and organoleptic tests and the housewife accepts it as such. To label such a product 'reconstituted' simply because you elect to make FCOJ even more convenient for her to use by performing the dilution for her is, to our way of thinking, unnecessary and possibly confusing." The end of the letter returned to why the word "reconstituted" might be confusing: "The housewife or dairy redistributes or merely adds water to it to return it to its original state. He or she does not perform any such complicated manipulation as the word 'reconstituted' implies."

Beacham was shown the letter but said that MacDowell's conclusions "do not represent my views."[11] Nor did he believe they would have represented MacDowell's views at the time of the hearings. Beacham was convinced that MacDowell would have agreed that processing had reached a level of sophistication that put FCOJ, and the ready-to-serve single-strength juice that the industry made from it, into categories all their own. No one verified the opinion of the "housewife" MacDowell cited in defense of his position.

There is evidence that processors did not go out of their way to determine the housewife's opinion about their product labels. One pasteurized juice distributor, Hood Dairies, hired Market Research Corporation of the United States to conduct a survey regarding consumer understanding of pasteurization as applied to orange juice. The company sought hard facts to back its case that the word "pasteurized," if placed on orange juice cartons, would confuse. Stanley Payne, the survey's director, testified that the "great bulk" of respondents "had no understanding." But Brennan was skeptical. He gleaned from the survey results that many women had a pretty good idea about the benefits of pasteurization, the elimination of germs being one. Payne countered that an understanding of the effects of the process did not translate into a "clear" understanding of the process itself.[12]

This issue remained unresolved, as Brennan turned to a bigger question: Why didn't Payne ask consumers whether they thought "pasteurization" should be declared on pasteurized orange juice labels? Payne's answer was simple: "[It was] not an issue posed to me by Hood Dairies."[13] Had the FDA been proactive, it might have surveyed consumers for their response to the crucial question.

Brennan continued to pursue the question with another witness, Sunkist representative Horace Campbell. Noting that the word "pasteurized" carried positive connotations of safety and shelf stability, Brennan posited that it "might be good to point that out to the consumer, would it not?" Campbell disagreed: "Not necessarily, she wouldn't understand it anyhow."[14] This position shielded processors from having to answer to the enlightened consumer who might want to try and hold the industry accountable for its practices.

The opposing testimony of consumer representative Faith Fenton and vice president of Tropicana David Hamrick captures the core of the disagreement between the FDA and the industry over what name to give pasteurized orange juice. Fenton insisted that orange juice labels should contain sufficient information for consumers to choose between "orange juice and orange juice products."[15] The distinction she made between juice and juice products indicated that she, like the FDA, saw orange juice as distinct and the progenitor of a family of processed orange juice products. For Hamrick, the process of concentrating orange juice produced "an extension of the product that you know as merely orange juice."[16] To him orange juice was a single entity with an ever-evolving identity.

The debate remains undecided to this day. The real issues—whether processing produces alien foods, and, if so, the public's right to know—are rarely addressed directly.

The orange juice standard of identity hearings were different from typical regulatory activity in that they penetrated the heart of the debate. The FDA determined that processed foods are indeed foreigners that require some government overseeing, adopting standardization as an appropriate mechanism for regulating food; and it agreed that consumers should be at least somewhat informed about fabricated foods, holding that "pasteurized" should be a part of product name if part of product process. At the same time, the agency didn't think that consumers should be informed enough to cause concern or confusion. This paternalistic policy is reflected in the lack of ingredient lists that for decades characterized the labels of standardized foods.

The cultural environment during the 1960s no doubt influenced the FDA's choice of standardization as the preferred method of orange juice regulation. As the hearings showed, it was a society in which the epithet "housewife" was used freely and uncritically; this sobriquet carried the assumption that the orange juice buyer was a woman who could not handle more than simple homemaking chores. The protective bent of standardization made sense for supervision of the supermarket sphere where the housewife, who was generally perceived as unable to comprehend scientific procedures, was sovereign. Standardization was especially fitting for orange juice, a product marketed into the 1980s almost exclusively to the all-American mother. The credentials of Anita Bryant, who was one of the most prominent orange juice promoters during this time, according to FCOJ inventor Edwin Moore, were that "she'd been Miss Oklahoma and I think she was runner up in Miss America."[17] The former beauty queen was frequently portrayed standing over the kitchen table, carafe of orange juice in hand, ready to serve her man and eager clan. The perfect hostess, she welcomed all of America to "come home to the Florida Sunshine Tree."

The surveys that Stanley Payne oversaw reinforced Bryant's advertisements: the lady of the house was in charge of the orange juice purchase. One survey that he conducted for the National Orange Juice Association sought to determine consumers' definition of orange juice. Explaining his method, Payne stated that one hundred women were interviewed at each of four malls in Los Angeles and New York. Although today this would stand out as a nonrandom selection, at the time it was a given that the orange juice buyer was a woman.

Under questioning from a Tropicana lawyer, Payne noted that he was not an authority on juice issues. Despite his intimate involvement in market research, he said he did not pay attention to advertising and therefore was not as observant in this regard as the housewife. His reasoning: "It is not necessary to my business. It is very necessary to her shopping, she thinks."[18] His response highlighted the gender divide. Food-related fanfare was the feminine domain, where the housewife was more expert than Payne.

The portrait of the orange juice buyer as a woman who could not grasp complex processing procedures reflected an attitude that had solidified among marketers by the 1930s. In his history of American advertising, Roland Marchand writes about the chauvinism of the advertisers of the 1920s, '30s, and '40s, who believed that "the typical consumer lacked their sophistication, suffered from daily anxieties about a number of minor product choices, and—most deplorable of all—was likely to be a woman." A 1928 A&P ad in the *Saturday Evening Post* presented two pictures, one of a business executive sitting in his office, and the other of a mother sitting with her daughter. The boldface caption at the top read, "Alone . . . she faces this problem." According to Marchand's interpretation, the ad "compared the businessman's wealth of expert advisers in decision-making with the lonely ineptitude of his wife as she faced the task of selecting food." Marchand also describes a

1933 ad in which the pharmaceutical firm Parke, Davis, and Company asked: "Does Mother *really* know best?" Marchand quotes social commentator Kathryn Weibel, who calls attention to the ad's insinuation that women needed authority figures to direct them "in the most mundane aspects of buying life."[19]

The hearings into developing standards of identity for orange juice and orange juice products indicate that society's view of women had not changed substantially by the 1960s. The outcome of the hearings—processed orange juice products with uninformative labels—reveals a point of agreement between the FDA and processor: "she" was simple, so best keep product labels as plain as possible. The elaborate contents and methods of processing orange juice still remain mostly under standardization's concealing cover. The mystery that continues to surround orange juice's production is perhaps a relic of the time when the food buyer was deemed to be a woman incapable of processing too much information.

VII

Regulating Misleading Orange Juice Labeling

I n addition to the labeling of pasteurized orange juice, the pervasiveness of confusing orange juice advertisements also spurred FDA standardization. The testimony of consumer representatives Faith Fenton and Anne Draper corroborated the FDA's suspicion that orange juice marketing was particularly prone to confusing the consumer. During her cross-examination by Edward Williams, a lawyer for the Florida Canners' Association, Fenton did not back away from criticizing the constituency he represented, offering testimony that was damaging to the juice manufacturers he defended: "One professor wrote that there was so much confusion about the labeling of orange juice products that she had for some time recommended that people 'eat oranges in their orange package'... that is whole oranges, to avoid confusion in the labeling of orange products." Evidently the tendency of the food industry to hide its products within ever more elaborate packaging mystified even those schooled in the field of food. Orange

juice labels had become so opaque that the experts trusted
only nature's wrapping.

The caution of the anonymous professor sent the message
that not only the content but also the container of orange juice
demanded FDA attention. For the contemporary reader it also
highlights the new occupations that evolving food-processing
technologies were creating for women. At the same time that
the processing sector was dispossessing women of their traditional status as primary provider, it opened for them the empowering role of food educator. In a world where food originated prepackaged in supermarket aisles, this was a role that
the average "housewife" could no longer perform without training. The modern food environment turned the former household occupation of nutritionist into a professional one that
was filled mostly by women. The home economists of the early
twentieth century pioneered the new profession and paved the
way for women like Fenton and her orange-juice-wary colleague to climb the career ladder. Because of her formal education in this area, Fenton could be welcomed into the public
sphere, where women had historically made mostly silent ap-
pearances in supporting roles. It brought her to Washington,
where she had the opportunity to voice her views before fellow
professionals.

The conclusion Fenton drew from her colleague's advice
signaled her success in infiltrating a predominantly male domain. She urged: "We don't want to see that become very general."[1] By "that," she presumably meant promoting fresh over
processed oranges. As such, Fenton's display of sympathy for
the industry in not wanting to turn consumers away from
processed orange juice demonstrated her capacity to alternate
between her roles as industry cohort and consumer confidant.
Although FDA lawyer Bruce Brennan and Sunkist's Horace

Campbell awkwardly all but donned the housewife's apron during the hearings to understand her juice perceptions and preparation, Fenton, dressed for all occasions, moved seamlessly between House and home.[2] Hers was an essential talent for any woman who wanted to be taken seriously beyond the threshold of domicile door.

Other consumer representatives who attended the hearings were not so well received. American Federation of Labor representative Anne Draper supplemented Fenton's professional opinions with the critical reflections of the everyday consumer. Her description of typical supermarket displays in which orange juice and fresh milk appeared side by side in similar cartons gave crucial context to accusations, such as the one made by the professor Fenton quoted, that orange juice labeling was confusing. Draper warned that such orange juice packaging and placement encouraged the consumer to falsely assume fresh unless otherwise stated.[3] She also identified the "reverse English" of the orange juice label, noting that the only way to know whether orange juice was freshly squeezed was if the label read, "100 percent pure, undiluted, untreated, unsweetened, freshly squeezed whole orange juice."[4] The message from Fenton and Draper was consistent: advertising jargon prevented consumers from knowing what they were buying.

Draper's testimony, however, incited little reaction, as her cross-examination by FDA and orange juice industry officials consisted of three short comments. The brevity of Draper's appearance begs the question of whether the time allotted to her was proportional to the esteem in which women who did not carry the rank of professor were held outside the domestic sphere. Regardless of how little discussion Draper's affidavit generated at the time, its relevance to the issues raised during the hearings makes it stand out among the hearings' more than three thousand pages of transcripts. Draper's testimony, combined with that of Fenton, left no doubt that orange juice marketing as well as manufacture needed regulation.

The testimony of Fenton and Draper strengthened the FDA's argument against the processor's claim that label changes would confuse the consumer. The mirror the two held to marketing convention reflected the consumer confusion that processors said they were seeking to prevent. The FDA's crusade against customs that confounded the consumer continued when Bruce Brennan cross-examined Glenco Citrus Products' Houston Kier about Glenco's pasteurized orange juice. Kier admitted that until recently his company was calling its heated-then-chilled, concentrated, and sugar-added product "fresh." The obvious discrepancy between product manufacture and marketing led Brennan to dig deeper into Glenco's historical method of juice labeling and distribution.

Brennan was curious why Glenco changed the name of its pasteurized juice from "fresh orange juice" to plain "orange juice." Kier replied: "Milton Duffy of the State of California frowned on the use of the word fresh in anything except in discussion of a cow." The undertone of Kier's response hinted at Glenco's objection to government involvement in orange juice labeling. Brennan drew his own conclusions and surmised that Glenco's stance was that, while abiding by the prohibition against calling its pasteurized product "fresh orange juice," it should still be allowed to call it "orange juice." He further posited the company's reasoning: the public, through company marketing, had come to know its pasteurized juice as "orange juice."[5] Kier confirmed that Brennan had guessed correctly.

Brennan then asked whether, if Glenco were still marketing its pasteurized product as "fresh" orange juice, it should be allowed to continue to do so. If Kier answered affirmatively, his response would stress the senselessness of a defense that condoned product mislabeling under the pretext of protecting good lines of consumer communication. A no answer would be an admission that consumer familiarity did not justify the use of inaccurate product names. Kier evaded the trap with an "I do not know."[6] As a result, the question of whether product labels should be able to remain as is, regardless of accuracy, was left unanswered.

Brennan's cross-examination of Kier launched the FDA's review of a series of marketing practices that it deemed detrimental to the interest of orange juice consumers. The deliberate lack of specificity with respect to Glenco's orange juice label was a benign ploy compared with how some of the larger companies were using the label for product promotion. One orange juice manufacturer, Parmet, marketed one type of orange juice under fourteen different labels.[7] Tropicana, then as now the most popular pasteurized orange juice producer in the United States, was among the shrewdest. FDA inspector Robert Lloyd Tillson dissected the label of Tropicana's sugar-added orange juice. Large type drew the consumer's eye to "Tropicana, Florida Orange Juice." Tillson noted that the issue of the wording was the same as with Glenco's product. "Orange juice," which was written in large letters, implied the product was fresh, not pasteurized. Tillson then pointed to smaller type on the label stating, "sucrose added," and remarked that "sucrose" was not the commonly used name for sugar.[8] Tropicana had managed to make a familiar sweetener foreign to the average consumer.

The cross-examination of Tropicana's vice president, David Hamrick, by Vincent Kleinfeld, lawyer for the National Orange Juice Association, introduced the thought processes of the minds promoting the common perception of orange juice as wholesomeness concentrated in a glass. Kleinfeld's interrogation of Hamrick also demonstrated divisions within the orange juice manufacturing community, possibly because of the competitive advantage that Kleinfeld later argued Tropicana achieved by not having to disclose its use of frozen juice in its pasteurized product.[9] Whatever the reason, this divisiveness added further clout to the FDA's bid to regulate orange juice.

Kleinfeld dug up more evidence of questionable orange juice marketing tactics than did all of the FDA officials and consumer witnesses combined. He focused on a particular label that told a tall tale about Tropicana's ready-to-serve pasteurized juice: "Tropicana 100 percent pure orange juice is squeezed from prime tree-ripened Florida oranges at our Cocoa Florida plant located in the Indian River Section, famous for its high quality oranges. . . . Tropicana orange juice is transported under constant refrigeration in stainless steel tanks by steamship to New York City. There it is packed in cartons and delivered to dairies in our mechanically refrigerated trucks."[10] As part of Tropicana's aggressive effort to differentiate its product, it singled out as distinctive juice characteristics that were meaningless and/or misleading. For example, its description of its oranges as "tree-ripened" was guilty of both. The adjective has meaning for fruits such as tomatoes that continue to ripen postharvest, but it carries none for oranges, which stop maturing once picked from the tree. All "ripe" oranges, and the Florida Citrus Code has long required that Florida juice be made only from mature oranges, were and are "tree-ripened."[11] Used in a paragraph in which space and cost restraints would have demanded that each word be carefully chosen, the adjective must have filled an important company purpose: no doubt to encourage the orange juice consumer to mistakenly assume that the phrase had bearing on orange juice quality.

In 1991, then–FDA Commissioner David Kessler declared that "'descriptors'—whatever appears on the label—must be accurate . . . the bottom line is that words on the food label must have meaning."[12] "Tree-ripened" fell below this baseline of acceptability. In 1961 Kleinfeld did not remark on the redundancy of "tree-ripened" and the impressions it might have generated. His questions centered first on what was missing from the label's description, and then on existing but inaccurate statements. With evident sarcasm, Kleinfeld asked Hamrick why, given the level of detail of the Tropicana label, Tropicana did not also mention that "the product has been frozen, stored for extensive periods of time and thawed." Hamrick responded that Tropicana did not "want to put anything on the label that would give the housewife the impression that the product is made from concentrated orange juice." Kleinfeld was quizzical: "Then the reason you don't want to make that statement [is because] you feel a truthful statement would mislead the consumer?" He was implicitly accusing Tropicana of building its version of truth on misleading impressions created by partial information, while using the risk of confusion as reason for not revealing all. Hamrick did not back away: "I'm quite sure it would."[13]

Kleinfeld's second line of questioning addressed the misstated rather than unstated. The label read that Tropicana's juice was squeezed in its Cocoa plant in Indian River, a region noted for its quality oranges, and he wondered whether Tropicana always used Indian River fruit. Hamrick answered carefully: "In the Cocoa operation . . . it is basically Indian River fruit," proceeding to admit, "I won't say every orange is Indian River." To this day the company's Bradenton plant in southwestern Florida processes most of Tropicana's oranges, and since juice plants tend to process oranges grown nearby, Hamrick more accurately could have noted that *most* of the oranges that Tropicana processed were *not* from Indian River. Tropicana's juice was and still is predominantly made from oranges from the unrenowned region of southwestern Florida, while Indian River is in northeast Florida.[14]

Hamrick refused to acknowledge that Tropicana's juice label distorted reality. When Kleinfeld asked whether Tropicana used Indian River fruit only, Hamrick answered rhetorically, "Is that what the label states?" Indeed, as Hamrick concluded, the label technically "is truthful . . . [the] claim is [that] the plant is located in the Indian River area." But Kleinfeld persisted: "Are you saying that the impression is not to convey to the consumer that the product comes from Indian River oranges?"[15] Here Hamrick's lawyer objected to Kleinfeld's attempt to establish Hamrick's personal opinion, and Kleinfeld's question was left unanswered.

When his cross-examination continued the next day, Kleinfeld continued his efforts to adumbrate shades of meaning as he pursued the question of the integrity of Tropicana's advertising. He asked: "Wouldn't the ordinary consumer, considering [the] statements made on [the] carton such as [']100 percent pure orange juice from prime tree-ripened Florida oranges at our Cocoa, Florida, plant, transported under constant refrigeration in stainless steel tanks by steamship to New York,['] think this product in this carton had been freshly squeezed and rushed for consumption up North?" He continued: "[You have] no opinion as to the impression currently conveyed, yet [you] believe a truthful statement [regarding] freezing would mislead?" No objection was raised this time, and Hamrick responded, "Correct. In a truthful statement any

design or impression which could be misleading to the consumer should have no place on the label." This principle, he reassured, "is our philosophy."[16]

Kleinfeld then turned to another Tropicana label that
claimed the company used the "fresh juice of fourteen tree-
ripened Valencia oranges" in every quart. Hamrick said that
the company contracted with an advertising agency that had
issued the label without authorization. He said that he did not
consider the label in keeping with the company's advertising
policy and ordered it withheld from further distribution. Kleinfeld persisted. He asked whether Valencia oranges were available from October through January. Hamrick admitted, "As a
rule, no." Kleinfeld quoted the rest of the label—"First fresh
orange juice with unlimited shelf life"—and then asked, "Is
this fresh orange juice?" Hamrick granted it was "not fresh
orange juice according to my definition." Should this therefore
be considered deceptive advertising? Kleinfeld asked. Hamrick
wavered: "[I] wouldn't say it in those words." He agreed the
label could be considered misleading but said, "I don't believe
there is any intention for deception." He proceeded to defend
the advertising agency, stating that it adopted "this method" of
advertising after taking note of the custom in the chilled juice
industry.[17]

Hamrick not only denied prevarication on the part of
Tropicana, he underscored his and, by association, Tropicana's
principles: "I informed them that we were not in sympathy
with the general method and wanted to stick to the facts in advertising our product." Two generalities in his statement weakened his effort at exoneration. First, Hamrick distinguished
between an unspecified "them" and "we." Presumably he intended "we" to refer to Tropicana and "them" to the advertising agency. But the division between the two could not have

been so clean. Someone at Tropicana hired the agency and must have approved the label's mass distribution. Because the decision regarding the type and placement of labels on a national product is not usually the work of one, *many* at Tropicana must also have been involved. Without more detail about the culpable parties Hamrick could not convincingly pass the ad off as the work of others. Second, Hamrick disassociated himself and his company from the censured label by testifying that the elusive but inclusive "we" did not sympathize with the label's "general method." Perhaps fearful of self- and company incrimination, he did not specify the "method" that called for his and Tropicana's disassociation. Hamrick's pronouncement that "we . . . wanted to stick to the facts" implied only that the "general method" strayed from the facts. How far Hamrick did not say.

The imprecise language that filled Hamrick's final attempt to clear his company of marketing misconduct was consistent with his performance throughout the Kleinfeld-led cross-examination. He spent two days dodging the single most important issue at stake: the degree to which untruthful ads tainted the trustworthiness of Tropicana, a trend-setting juice processor.

The evidence of deceptive and confusing orange juice advertising unearthed at the hearings was more than sufficient to justify FDA standardization of orange juice. The FDA's biggest challenge was not to defend its decision to standardize but to determine how best to do so. The following chapter looks at the FDA's solicitation of the processor's input to "fix," according to s.341 of the Federal Food, Drug, and Cosmetic Act, standards of identity. To determine what to include in the standard of identity for each type of orange juice under consideration, the hearings' participants focused on the following: (1) What are the constituents, processes, and additives that affect an orange juice product's identity?; (2) How should "identity items" be fixed in a standard of identity?; and (3) How much should the label say about the product? These three questions were united by a broader one: What does the consumer expect orange juice to be? The next chapters reveal that there were no easy answers.

VIII
Regulating Content

Over the course of the hearings processors and regulators spent untold hours trying to isolate the essential components of the various kinds of orange juice on the market. The exercise emphasized the elusiveness of orange juice's identity, in whatever form it took. First up for examination was fresh orange juice. Even this seemingly straightforward juice was perplexing. Horace Campbell of Sunkist said, "Freshly squeezed orange juice . . . is a widely variable commodity. . . . I don't know just how you would describe it."[1] The inconsistency that is inherent in a fresh product is a headache for those seeking to define it.

One of the challenges was to separate variations that were critical aspects of orange juice and therefore "identity items"—such as natural orange sugar—from those that were not. After Campbell noted the ranges in orange oil, seed content, pulp levels, and other components that made freshly squeezed orange juice a variable product, Edward Williams, a lawyer for the National Association of Frozen Food Packers and the Florida Canners' Association, put the crucial question to Campbell: "Do any of the variations in composition which you discussed . . . change the product 'orange juice' from orange juice to a product of some other nature or some other name?"[2] Campbell's answer was practical: the standard of identity for fresh orange juice should set an appropriate sugar-to-acid ratio and percentage of orange sugar solids; it shouldn't define pulp and seed content. Manufacturers of processed orange juice would balk at a standard for fresh orange juice that defined the juice as seedless, because having to limit seed content for their processed product too would slow down production lines.

Ultimately, however, the real debate that took place between regulators and juice makers was not about defining the material components of identity but the processes and additives that might change it. Chemical and physical composition became, for all parties present, the key signifiers of whether a process or additive transformed a particular product's fundamental nature. The effects of freezing and adding water were analyzed especially rigorously.

Processors parted ways over whether freezing orange juice changed the juice's identity. John T. R. Nickerson, an associate professor of food processing at Massachusetts Institute of Technology and a witness for the National Orange Juice Association, explained that freezing altered orange juice because some substances (such as flavor-providing chemicals) evaporated and were therefore lost from juice that had been frozen at different temperatures and for different durations and then defrosted. He concluded that "the product before and after freezing is not identical."[3]

Tropicana vice president David Hamrick disagreed. He held that "in freezing there is no addition or subtraction of volatile components or any change in chemical characteristics." In his view freezing was "merely a degree of refrigeration." According to Hamrick, science showed that the process imposed did not alter the chemistry of the juice. Although the two disagreed over the results, they thus implicitly agreed that chemical composition was an appropriate basis for making a decision regarding identity-reshaping activity. Hamrick and the others who testified that freezing was inconsequential must have been convincing.[4] After the hearings almost all the major pasteurized orange juice manufacturers began adding frozen to their heated juice to ensure juice stocks would be high when supplies of fresh oranges were low. Tropicana still uses frozen orange juice in a product that complies with the FDA's standard of identity for pasteurized orange juice.[5]

The influence on juice identity of the addition of water also brought science into play. Robert Kilburn of the Florida Citrus Canners' Cooperative relied on chemistry to justify the many routes by which water is introduced into frozen concentrated juice. He described for FDA lawyer Bruce Brennan the beginning part of the procedure for adding fresh juice to frozen concentrate—a critical step for making the final product palatable. First, processors removed from the freezer the fifty-five-gallon drums in which they stored concentrate and allowed the contents to soften before transferring to a hopper for mixing. About a gallon of concentrate clung to the interior, which processors flushed out with water into the hopper. Kilburn estimated that prohibiting the practice would mean that the Florida orange juice industry would lose 250,000 gallons of concentrate every year, worth about three dollars per gallon, and this cost would then be added to the retail price of concentrate.[6]

Kilburn rejected Brennan's suggestion that the addition of this "foreign" water tainted the product and altered its identity. He reasoned that "fundamentally we are selling soluble fruit solids to the consumer when they purchase frozen concentrate. . . . The article of value which is sold to them is the soluble solids."[7] To Kilburn, frozen concentrate was a specified amount of orange sugar solids. Consequently, the introduction of water was not, in his opinion, a problem because it did not dilute the percentage of sugar solids in the product; he insisted that processors either evaporated the extra water away or compensated by adding more concentrate.

Brennan was not convinced that the consumer's sole expectation of frozen concentrate was a specific quantity of concentrated sugar. He asked why, if consumers considered frozen concentrate to be merely a can of orange sugar, they often referred to it as frozen orange "juice."[8] Brennan seemed to be implying that consumers did consider frozen concentrate to be more than sugar, and that if they knew that what they called juice had been concentrated, watered down, and then reconcentrated, they would be disappointed. Kilburn disregarded common usage as a gauge of the consumer's expectations. He traced the phrase "frozen orange juice" to the early days of frozen concentrate marketing and reminded Brennan that the label on the can read frozen "concentrated" orange juice. Brennan was still not persuaded; if processors could play with the sugar level of concentrate by adding and evaporating water, he saw no reason why the same shouldn't apply to orange juice canning. Based on Kilburn's testimony, Brennan reasoned that canners should be able to add water to highly concentrated juice to bring the sugar down to an acceptable level. Kilburn said the situation was different for canned juice because adding water to it "would not conform to what most people understand to be orange juice. . . . Canned juice is not diluted." Although the sugar solids of canned juice vary naturally, he said, "uniformity is a very distinguishing characteristic in concentrated orange juice, and the consumer expects it to be very consistent in composition."[9] Kilburn insinuated that as long as all the elements of concentrate comported with concentrate's overall formula, the consumer would be satisfied; the processes and substances used to get there were incidental to its identity so long as the final product contained the acceptable amount of solids.

The FDA's Lowrie Beacham did not condone chemistry as the measure of a juice's identity. If chemistry was the critical consideration, he pointed out, fresh, pasteurized, and reconstituted juice—all having almost identical chemical makeups—would share the same identity. He contended that "the basis of differentiating a product is not chemical composition or taste but manufacturing process."[10] To illustrate his point he compared pasteurized orange juice, which has never been concentrated, and reconstituted juice, which is diluted from concentrate. With reconstituted juice, he said, "water is removed in Florida, or wherever, and the water added back is entirely different water. . . . So that you do not have the same identity, even though the physical, chemical characteristics are so closely one to the other it is difficult to distinguish between them." The two-step procedure for making reconstituted juice—juice was concentrated in Florida and then transported to the point of sale where it was reconstituted with water—spoke for itself: "Knowledge of the facts, now common sense tells us that we do not have the same product. We have a different product."[11]

Kilburn's assumption that FCOJ consumers cared only about final composition gave the green light to shortcuts that some might consider dubious. Even if Kilburn was correct that consumers saw concentrated orange sugar when they looked at a can of FCOJ, his willingness to give the consumer the final word on product acceptability was problematic given his awareness of how little the consumer knew about specifics. He recognized, for instance, that the consumer did not know that orange oil was a constituent. Acknowledging that flavor and appearance were the consumer's only means of appraisal, he concluded that "as long as those two factors are satisfactory, she assumes composition to be correct."[12]

Kilburn's observation weakened his position that uniformity was sufficient to make frozen concentrate consumers happy. It also exposed the impaired judgment of consumers, whose approval Kilburn took for granted in justifying the common practice of adding "foreign" water to frozen concentrate. Processors could easily manipulate flavor and appearance to the consumers' liking using ingredients and procedures that would not necessarily meet with the consumers' approval. A juice made partly with cane sugar could be made to taste and appear the same as one that was 100 percent juice. Consumers, whom Kilburn admitted relied wholly on sight and taste to evaluate juice, could thus easily mistake one for the other. Their limited juice assessment abilities made them unreliable judges of acceptability.

Kilburn not only was mistaken in making consumer approval of the final juice the test of acceptability of a particular processing procedure. He was also incorrect in assuming consumers held to a "correct composition" for frozen concentrate that could be determined by flavor and appearance alone. One witness, Ruth Robinson, who testified on behalf of the Federation of Homemakers "to ensure that no artificial flavors, colors, or preservatives are permitted in our orange juice products," revealed that consumers were not so easily satisfied.[13] Apparently the homemaker expected more from a can of FCOJ than a set amount of orange sugar solids. She wanted a pure product she could feel good about serving her family. Robinson's appeal to regulators suggested that homemakers did not feel confident that their senses alone could decipher between desirable and undesirable juice. Contrary to Kilburn's testimony, the homemakers Robinson represented might, had they known, have also disapproved of the common practice of washing out barrels of concentrate with water that was then reused to make concentrate. A uniform orange juice with a satisfactory taste and appearance did not alone make for a satisfactory orange juice.

If consumers were more demanding than Kilburn and his colleagues portrayed them to be, they were also less rigid. Kilburn introduced the idea that the consumer had a fixed notion of FCOJ. He was not the last to do so. James Hodges, president of a California-based processor, assumed the same in addressing where to set the base level of orange sugar solids for reconstituted juice. The decision had been made that orange sugar solids were an "identity item" of reconstituted frozen concentrate. The next issue for deliberation was the appropriate amount. Hodges had a number in mind based on what he imagined went on in the kitchen. He justified 11.3 degrees Brix (Brix being the percentage, by weight, of sugar in a solution at a specific temperature) on the grounds that "when the housewife prepares juice from a six-ounce case or can of concentrate, her instructions are normally to add three parts of water to one part of concentrate." In his scenario the "housewife" followed the can's instructions exactly: "She, of course, does this by filling the can three times and pouring the water in with the concentrate."

Keeping to his image of an aproned robot reconstituting juice with emptied cans of concentrate, Hodges explained why the FDA's proposed 11.8 degrees Brix was too high: "Now, there is a small but necessary headspace when the frozen concentrate is in the can. When the housewife fills the can with water, she does not allow for this headspace. Therefore she is actually adding slightly more than three parts of water, and the product which she serves her family therefore is actually about 11.3 degrees Brix, rather than 11.8 degrees Brix."[14] Assuming that the goal was to replicate what the housewife produced from concentrate, he did not believe that the FDA's calculation accounted for the water that the housewife unintentionally but invariably added to each can of concentrate over and above the can's three-to-one directions.

The flaw in Hodges's reasoning was his premise that the housewife always followed the processor's orders. Renée Goodrich, a food science specialist at the Citrus Research and Education Center in Florida, grew up on FCOJ and recalled that "my mom used to add a lot of water back to the FCOJ to stretch it."[15] Like mothers everywhere, Goodrich's did not strictly follow the directions on the can; she had her own idea as to how to reconstitute juice; for her, homemade reconstituted orange juice varied according to the circumstances. On a hot day she might add more water so that the juice would go further. On special occasions she might use less water to make the juice stronger. Hodges failed to consider that she was not simply another arm of the industry, nor did she speak the processor's language fluently. Processors could therefore not program her completely; she moved independently, and each time a little differently. The evidence indicates that the attempt to "fix" standards of identity on the basis of the consumer's expectation of uniformity was an exercise in futility.

Frank J. Portras, a member of Plymouth Citrus Products Cooperative, could have been speaking on behalf of every processor when he volunteered, "We at Plymouth as well as the rest of the industry wish Mrs. Housewife to have the best and the most uniform product possible which would promote honesty and fair dealings in the interest of customers." William Risteau, the hearings' presiding examiner, asked Portras, one of the hearings' first witnesses, whether the processor's objective was to make a product to "a standard which is as uniform as you can get it, and which grades well by the [quality] standards of the USDA?" Portras's "yes" response established the processor's organizing principle. Good-quality juice was not sufficient. Marvin Walker, general manager of Florida Citrus Canners' Cooperative, confirmed that processors were in the business of "seeing to it that the consumer of the United States receive products of . . . not only good quality, but uniform quality."[16]

Processors defended the priority of uniformity differently. At times they said that's what consumers wanted, at other times they equated it with quality. In his first appearance at the hearings Kilburn asserted: "Uniformity is an essential characteristic of good quality. . . . Several processing techniques have been developed specifically to level out, as much as practical, the variability in oranges." Later in the hearings he repeated: "[We] tend to think of uniformity being somewhat synonymous with quality."[17]

The processor's goal of uniformity did not go unchallenged. Risteau asked Portras whether consumer tastes were uniform with respect to the orange juice they liked. Portras responded, "I don't think there is much difference in housewives." Seemingly lumping all housewives together, he said that if they liked the product, "they would like to have it uniform."[18] But the rebellious ways of Goodrich's mother, who produced a reconstitute that was unrecognizable to the processor, indicated that Portras needed to do more research before he could accurately state what the housewife wanted.

The FDA's suggested resolution to a particular problem

indicated that not even the agency always put the consumer's interests first in its drive for standardization. At the end of the Valencia orange season, which falls around June, the ratio of sugar to acid of the remaining oranges on the tree is too high for processing. Oranges with insufficient acid make for an insipid juice, so processors blend the overly ripe fruit with more acidic—that is, low sugar-to-acid ratio—juice. Minute Maid's Charles Brokaw explained that sometimes his company concentrated large quantities of juice during the early part of the season, when the sugar-to-acid ratio of the available oranges is low. It stored some of this concentrate for mixing with Valencias picked in June. But he said the procedure was not practical because of the difficulty of knowing how much of the juice would be needed later in the year.

Brokaw preferred another solution: combining fresh juice from *Citrus aurantium,* the sour orange, with late-season Valencia juice. This approach cut down on storage and required relatively little sour orange juice to raise the acid amount. However, Brokaw testified that processors did not always use sour orange to adjust high sugar-to-acid-ratio juice, because demand for sour oranges exceeded supply. According to Brokaw, Florida grew only 50 percent of the amount processors required. Moreover, he said, sour oranges cost more than low sugar-to-acid-ratio sweet oranges.[19]

Brennan was reluctant to allow sour orange juice into a product called orange juice. *Citrus aurantium* is genetically related to but is not the same fruit as *Citrus sinensis,* the sweet orange. The FDA was prepared to permit the use of a certain percentage of juice from *Citrus reticulata,* or the mandarin, to enhance color. Brokaw thought the FDA should define FCOJ's standard of identity to allow for the use of some sour orange juice, too. But Brennan wondered whether processors had ever considered using "other *Citrus sinensis* which might be available in June from a source outside of Florida." He asked specifically about using the higher-acid California Valencia. Brokaw said the availability of fruit in California was limited. He also called attention to the high cost of transportation and logistical difficulty of scheduling the oranges' timely arrival in Florida.[20] In the end, Brennan's proposal was discarded; transporting fruit from outside Florida would raise the price of orange juice, decrease its nutritional value by substituting travel-worn California fruit for fresh Florida oranges, increase energy consumption, and diminish the market for Florida's sweet and sour oranges. As Brokaw had proposed, the FDA standard of identity for FCOJ provides for the use of up to 10 percent *Citrus reticulata* and 5 percent *Citrus aurantium* juice.[21] Still, along the way Brennan had exposed the FDA's misguided fixation on uniformity. In the name of uniformity he suggested an idea that would be unfavorable to the consumer, the Florida farmer, and the environment.

Processors likely prioritized uniformity because it suited their mechanized methods of operation. Regulators did so because the Federal Food, Drug, and Cosmetic Act, via the s.341 standard of identity, implicitly told them to do so—achieving uniformity is what standardization is all about. Both did so despite the fact that witnesses such as consumer representative Ruth Robinson clearly indicated that the consumer wanted more from processed orange juice than constant composition.

In the final stages of the hearings Lowrie Beacham summarized what he knew about FCOJ:

> On the basis of the knowledge that I have of concentrated orange juice gained from observations in

plants where it is prepared, conferences with technical people, testimony that has been presented here, and other sources of knowledge, I would say that frozen concentrated orange juice is a highly sophisticated product in the sense that it is fabricated from various orange components and is manipulated to meet predetermined specifications.

It is not simply orange juice concentrated by evaporation and then frozen. Instead it consists of a basic blend of selected orange juices that are available at the time of manufacture, to which may be added frozen concentrate of other varieties, packed at other seasons and perhaps other places.

For this other frozen concentrate is usually added to get a desired ratio or desired color or perhaps both. Refined orange oil is added to the desired concentration of orange oil and we have now heard testimony on the addition of orange essence as well.

Course [sic] pulp is added to contribute pulpiness and this may be highly colored Valencia pulp, separated and frozen the year before to improve color.

The final divided pulp that may be present is reduced to whatever level is called for by various techniques including high speed centrifuge, fractions or all of the product may be heated before blending and washed pulp extract or second finisher liquid may be among the components used.

The resulting product, when properly diluted, makes a delicious beverage which resembles and substitutes for fresh orange juice, *but it has an identity all of its own* [emphasis added].[22]

His description illustrates why the identity of ostensibly simple products became controversial. FCOJ was no further removed from fresh orange juice than the rest of the orange juice interpretations on the table. Pasteurized, canned, reconstituted, and all the other processed juices each had "an identity all of its own." The difficulty was how to capture the essence of each one's complex persona.

Although processor and regulator strove to find an identity that would ensure product predictability, their approach was erratic. Sometimes they were solicitors, invoking the opinion of the consumer, albeit rarely directly. She would have been more of a diagnostic aid if they had asked her personally. Other times they acted as chemists, and still others as pragmatists. The role they played determined the resolutions they made, which were not always internally coherent. The task of defining orange juice standards of identity thus got messy. Things got especially sticky when processors and regulators delved into a process that many present at the hearings were hearing about for the first time: essence addition.

IX

Regulating the Essence
of Orange Juice

The hearings made the FDA aware that processors were experimenting with using orange essence in the manufacture of juice. The addition of essence and refined orange oil to products advertised as 100 percent pure orange juice added a new regulatory issue to the FDA's agenda: whether orange essence and oil should be considered additives rather than natural components of orange juice and, if so, whether they should be labeled. In the early 1960s "cut-back," the fresh juice that was put back into concentrated juice to reduce the Brix and add taste, was still the flavoring material of choice for FCOJ. Homer Hooks, general manager of the Florida Citrus Commission, singled out the invention as one of the most important achievements of the Florida Department of Citrus's research program, crediting it as being largely responsible for the immense popularity of FCOJ.[1]

In addition to cut-back, Howard Trumm, a specialist in citrus technology at the juice company Libby, McNeil, and

Libby, described a complementary practice at the hearings. During juice extraction processors typically separate out a portion of the fruit's orange oil, which Trumm said contained many of the flavoring constituents of orange juice. This oil was often combined with cut-back to restore the flavorful chemicals, or "volatiles," that evaporated when juice was boiled down to concentrate. At the time of the hearings Libby, McNeil, and Libby was one of a handful of companies at the forefront of flavor research, as it had begun experimental work on a commercial scale to recover not only orange oil but also essence. Trumm described orange essence generally as the volatile flavoring components that escaped from orange juice through evaporation and that were then recovered from the water vapor and concentrated as orange essence.[2]

(A Firmenich flavor engineer whom I interviewed helped clarify technical terms that Trumm and others used at the hearings. Although the oil from the peel of the fruit goes by more than one name, he explained, "when someone says orange oil, 98 percent of the time . . . they're talking about orange peel oil." To flavor engineers, he says, the substance is simply "orange oil." The orange oil that Trumm identified as a customary addition to concentrate likely was this by-product of the peel. After the orange is peeled, it is sectioned and its juice extracted. At this point two other by-products can result if the juice is evaporated for concentration: oil phase essence and water phase essence. Oil phase essence is usually called "oil essence," "essence oil," or "juice oil." Water phase essence also goes by the name "aqueous essence." Finally, the flavor engineer explained that essence oil and peel oil are both also considered "essential oils," defined by him as "any oil that can be extracted or cultivated from a natural product." Presumably the "orange essence" that Trumm and others referred to at the hearings was a catchall for oil and water phase essence. In the early days of juice processing the value of these essences went unrecognized, the flavor engineer said. "They were dumping them to the ground. They were pouring them [down] the drain. They didn't understand the value when they started up the concentration process. It was a few pioneers in the industry that went out there and said this is liquid gold."[3] Houston Kier of Glenco Citrus Products was one of those at the hearings who admitted he was unfamiliar with the term "orange essence.")[4]

Today's citrus processing specialists agree that the history of orange essence recovery has not been fully written. Renée Goodrich of the Citrus Research and Education Center (CREC) in Lake Alfred, Florida, said that oil and essence's replacement of cut-back as the preferred orange juice flavoring method "is just one of those kinds of things that everybody knows about but there's not a lot of documentation."[5] The scarcity of material written on the industry's transition from cut-back to the complex "flavor packs" now in use makes the depositions of a handful of witnesses at the 1961 hearings an invaluable resource for understanding this evolution in orange juice processing. The testimonies of Trumm and of Robert Shaffner, vice president of research and quality standards for Libby, McNeil, and Libby, and Richard Wolford, associate chemist for the Florida Citrus Commission, in particular provide unique sources of orange essence oral history.

Wolford was part of a joint effort between the Florida Department of Citrus (FDOC) and the juice processing sector to solve the mystery of fresh orange juice's flavor. Libby, McNeil, and Libby, which had spent years perfecting an essence recovery procedure, was the major supplier of essence for the essence-related experiments that FDOC citrus by-product researchers such as Wolford were performing from the CREC, where Goodrich now works. The objective of one project, filed as Florida Agricultural Experiment Station Project # 1033, was the "ultimate characterization of the flavor of fresh citrus juices and processed citrus products." As of 1961 the project had been officially on record for only one year. But Wolford noted that FDOC researchers were performing analyses of orange essence as far back as 1956, when the FDOC first had access to the instruments that separate substances into their constituent chemicals.[6]

At the hearings Wolford described the results of an analysis of orange essence. Forty different components were detected, though not all of them were identified. Identification was lagging, he said, not because the chemicals were trade secrets, but because essence fractionation and identification was "just now becoming a real known science." Further, because the technology was still developing, he allowed that there could be even more than the forty components already detected. He was correct to be conservative; CREC citrus by-product researchers estimate today that the number identified reaches into the several hundreds.[7]

Regardless of the crudeness of their diagnostic tools, the 1960s researchers were able to make some groundbreaking discoveries. They divided the volatiles they found into five major classes: carbonyls, alcohols, esters, volatile organic acids, and terpene hydrocarbons. Each, Wolford could say with certainty, contributed more or less to the characteristic flavor and aroma of freshly squeezed orange juice. He hypothesized that carbonyls such as aldehydes contributed the most.[8]

Essence experimentation was not restricted to CREC chemistry labs. Shaffner testified that the first orange juice pack on record to contain orange essence was marketed in 1959. Defending his position to a regulatory agency concerned about the addition of any substance to a product presented as pure, Shaffner noted that juice processors had been recovering essence from pineapple concentrate, strawberries, and raspberries well before 1950. Even apple juice processors were adding apple essence to their final product. He also pointed to the standard of identity for frozen desserts, which permitted the use of essence as an ingredient.[9]

The timing of the hearings is critical to the understanding of the development of not only the science but also the regulation of adding flavor back to products labeled 100 percent orange juice. Before considering *how* to regulate orange oil and essence, the FDA had to be convinced that the use of the two was common enough to deserve the agency's attention. Even Trumm, who spoke enthusiastically about the benefits of orange essence, showed some uncertainty about its potential. He volunteered that Libby, McNeil, and Libby never used orange essence alone in its experimental packs. It always combined essence with cold-pressed oil, otherwise known as orange oil. Sometimes it also added cut-back. Trumm did not dismiss the prospect of using essence on its own: "If essence can contribute as much flavor and aroma as cut-back, it may become possible to substitute essence for cut-back." His company had not yet decided which route was best for producing a "uniform [and] high quality product."

Wolford, who had been working closely with Libby, McNeil, and Libby, was more cautious. He was unsure whether the three methods for restoring flavor to concentrated orange juice—the addition of cut-back, cut-back plus orange oil, and cut-back plus orange essence—were mutually exclusive or complementary. Observing that the cut-back and orange oil combination had been used for years, he was prudent about orange essence's future: "I don't know whether essence alone can adequately restore lost volatiles but it appears that it will favorably contribute to the fresh aroma and bouquet when used with oil or both oil and cut-back."[10]

Cost was also a factor in predicting the degree to which processors would rely on essence as a flavoring material. Trumm testified that the essence recovery and restoration processes were "considerably more involved and expensive than the cut-back process." But Brennan pointed out in his cross-examination that essence promised savings for processors of frozen concentrate, noting that if processors added essence instead of cut-back they would not have to remove as much water in the evaporator. Trumm admitted that processors did "over-concentrate to allow for the addition of cut-back" and that this practice would not be as necessary if essence was substituted for cutback. The cost of recovering and making essence would, he ceded, "offset the lesser cost [of cut-back] by not evaporating as much." Further, Trumm noted, by removing the need for "over-concentrating," essence could "possibly improve the quality [of concentrated orange juice] by reducing the retention time of juice in the evaporators."[11] It was becoming clear why processors might choose essence over cut-back to restore fresh flavor to processed orange juice, and so the FDA decided to take essence addition seriously.

The debates that developed during the final days of the hearings over orange essence and oil introduction touched on a few fundamental issues that are still debated today: how to differentiate between artificial additives and natural ingredients; an additive's proper use versus its abuse; and essential education versus too much information.

The first witness at the hearings, a Mr. Townsend of the Cooperative League of the USA, was also the first to express

disapproval regarding essence and oil reintroduction. Representing a federation of consumers consisting of fourteen million U.S. families, he read a statement from home economist
Betsy Wood, who set out the group's uncompromising position: "We are not opposed to processing necessary to preserve
flavor or vitamin content but we do object to manipulations
whose sole object is to disguise an originally inferior juice.
Most of the proposed additives could be used for this purpose." Orange essence and oil were examples of "additives" the
organization deemed unacceptable: "Orange oil and orange
essence should not be permitted as optional ingredients. It is
our understanding that these ingredients would only be needed
when the product itself is of inferior quality."[12]

Wood's testimony reflected one of the FDA's biggest concerns regarding essence and oil addition. Later in the hearings,
after Brennan compelled Trumm to admit that the substitution
of essence for cut-back would save the processor money, Brennan asked: "If [essence] is so strong, [is it] strong enough to
cover unusual or off or cooked flavors that be in the juice?"
Trumm's concession that essence lessened the degree to which
juice needed to be concentrated had underscored essence's po-
tency and that it could be used to overpower the odors of bad
juice. Trumm answered, "Essence will not, it is not going to
cover up any basic defect you have in the juice."[13]

Trumm's colleague Shaffner explained why he thought
essence should be a permissible ingredient in the standard of
identity for frozen concentrated orange juice: "I think that if
we can supply the consumer with products that most closely
resemble the natural product, in this case orange juice, [and]
we believe that essence does go a long way to do that, that it is
in the interest of the consumer to include orange essence."[14]

Wolford's testimony reinforced Shaffner's. While recognizing "consumer acceptance of present orange juice is satisfactory," he expected that acceptance would improve "by restoring more completely the volatile flavoring components than is possible under present circumstances." Highlighting the boon essence would be to the consumer was strategic considering the FDA's consumer mandate.

Apparently skeptical of flavor scientists' concern for the consumer's interests, Brennan asked Wolford, "Have you in your experience received any complaints or information from consumers that they just as well not have the concentrated orange juice taste like fresh orange juice, but rather have it taste more like they know as concentrated orange juice?" Wolford's reply intimated surprise: "No I sure haven't." Brennan continued: "Some representative of a large concentrating concern has said that some of the younger generation actually not having association with fresh oranges nearly as much as previous generations, possibly hardly at all, some of the younger generation has expressed a dislike for the hearty orange flavor that one might get from a freshly expressed orange and has been said to prefer a less strong flavor." Brennan had evidence that processed orange juice had so dulled consumers' senses that they were becoming attuned to the taste of less-flavorful juice. He questioned whether North Americans desired, or would even find palatable, fresh squeezed.

At this point Roderick Shaw, an attorney representing the Florida Canners Association, demanded a page reference for Brennan's statement. Brennan downplayed the source, saying that he was interested in only whether consumers preferred the taste of processed over freshly squeezed juice. Wolford understood that Brennan seemed to be questioning whether the consumer's interest really directed processing research. He took over from Shaw: "We don't attempt to speak for the con-

sumer without factual information. Therefore, instead of trying to force the development of concentrate in any direction
on our own opinion, we rely on the opinion of consumers
through taste tests. Therefore, instead of trying to make the
concentrate taste like something that we think it should be, we
actually test a new method of production or a modification to
see if the consumer prefers this new method."

Brennan called Wolford to account for his pledge of re-
sponsiveness to consumer preferences: "Do you have taste tests,
results of taste tests available for us here to indicate that the
consumer would like to have more volatile flavoring constituents in her orange juice than is presently being obtained?"
Wolford confessed that he himself did not, but he referred to
an unnamed "other party," whom he said "has brought information of the results of such tests."[15]

Marvin Walker, general manager of Florida Citrus Canners' Cooperative, presented a different picture of how processors operated: "Consumer preferences are not determined by
getting a bunch of consumers around the table and passing out
samples of juice and saying 'which do you prefer?' They are determined by the purchasing agents principally of the large
chain stores in the United States." Modern-day food scientists
corroborate Walker's version of food retail reality. In response
to whether today's market for orange juice is consumer driven,
the CREC's Renée Goodrich said, "No, [the market is] probably retailer driven as much as anything."[16]

Walker then noted that the marketer, in turn, influenced
the purchasing agent: "Why does a purchaser purchase tanger-
ine juice, orange juice, tomato juice, pineapple juice? It is because
of the promotion behind it, the advertising behind it." The
consumer's "personal experience with the product," Walker
said, came second.[17] Wolford's depiction of the consumer hold-

ing a direct line of communication to the processor was not very convincing next to Walker's depiction of a consumer who had little say in how processed products were made.

The concerns that Townsend and the FDA voiced about essence's ability to disguise poor juice merged into related and equally contentious issues. Wolford testified that orange essence comprised "volatiles that naturally occur in freshly extracted juice." Anticipating FDA inquiries about whether recovered essence was somehow different, he went on to say that "significant chemical changes" did not occur during the recovery process "because no extraneous materials are permitted to come in contact with the volatiles, and the temperatures and pressure used in recovery are within a range at which no significant chemical change would occur." According to Wolford, there was "no reason why [essence] shouldn't be permitted to be restored." Robert Kilburn of the Florida Citrus Canners' Cooperative used similar language to defend essence restoration: "Essence is composed entirely of the natural constituents of orange juice."[18] Their statements insinuated that processors should be able to add back to their product substances that escaped during processing if the restored substance and final product looked like their unprocessed relatives.

Brennan doubted the premise. He was unconvinced that adding orange essence to orange juice did not create a different product. He asked Trumm to identify the criteria that Libby, McNeil, and Libby used to "ensure that you don't use more essence in concentrate than might have been in single strength juice." The line of inquiry suggested that FDA approval turned in part on proof that essence's use would not create juice with a substantively different flavor profile than that of fresh juice. Trumm responded vaguely, "Through various determinations and calculations."[19]

Even more troubling to the FDA was the prospect of processors using imitation orange essence in their juice. Asked by Brennan about the composition of artificial flavoring, Wolford responded that such formulas were a "closed secret among the formulators," but acknowledged that the flavoring material normally used was composed of chemicals that could be produced synthetically in a laboratory.[20]

Lowrie Beacham, deputy director of the FDA's Division of Food, likewise expressed his concern about synthetic reproduction. Ed Williams, lawyer for the National Association of Frozen Food Packers, asked him during cross-examination: "Does the reason you are not in favor of the addition of essence have anything do with the fact that it is possible to make essence synthetically?" Beacham conceded: "That, very probably, is in the background of my objections, that adding essence opens the door to the addition of synthetic essence as well as those that are naturally recovered." Williams asked whether the FDA would be able to detect synthetic essence in orange juice. "It depends on the nature of the synthetic essence," Beacham answered, adding that it would be a "difficult undertaking." In earlier testimony he was more specific about the FDA's ability to distinguish "actual" orange essence from artificial flavoring: "It would be difficult to determine by an objective examination . . . a chemical examination of a sample of the finished product. . . . [It] might not be difficult to determine by means of factory inspection and other investigative techniques." In other words, the FDA would have to use the power of observation rather than chemistry to catch the use of synthetic flavor.

Williams pointed Beacham to the fact that the FDA's incapacity to differentiate between natural and artificially derived ascorbic acid did not stop the agency from allowing that nutrient in orange juice. "Is there any better reason for exclud-

cent orange juice products, a senior employee there smiled coyly and said, "I would have to shoot you if I told you."[27]

The culture of secrecy surrounding the manufacture of orange juice flavoring ingredients did not sit well with the FDA in the 1960s. After the juice processors' consistent dodging of Brennan's inquiries, Brennan cornered Trumm: "Are you asking the Secretary to permit the addition to orange concentrate, something that you don't know what it is, you won't tell us how you make it?"[28] The question was left unanswered. As the hearings progressed, FDA examiners relaxed their effort to break down the wall of secrecy that protected the processors' experimentation with orange essence from public scrutiny, perhaps in recognition of the futility of their attempts.

As processors fought to allow the addition of essence to orange juice, they also negotiated where to set the limits. Processors and regulators agreed that the amount should not exceed that which was naturally found in juice. The next issue was to define the types of orange juice to which essence could be added. Beacham observed that "adding essence is in the nature of restoring something that has been lost," and concluded that essence could legitimately be added only to concentrate. He elaborated: "Where concentrate is not used I see no reason for its use. In a single strength juice where no concentrate has been added, the essence has not been lost and I see no need for restoring something that has not been lost." There was thus no need, he said, for "the addition of essence to pasteurized orange juice where concentrate is not added."[29] Beacham's resolution rested on the premise that processors should be able to reintroduce into their juice only natural substances that disappeared during processing.

Using Beacham's logic to challenge the limitation of essence to concentrate, Trumm explained that deaeration, a procedure that some pasteurized orange juice processors used to prevent their juice from oxidizing in storage, "removes lots of volatile components." Since volatile components were lost, they should, he said, applying Beacham's reasoning, be able to be returned.[30] But the FDA's reluctance to allow essence in pasteurized juice was more complicated than that. The fact that essence is largely water created a problem. Brennan wondered whether Trumm was saying the FDA should allow essence in pasteurized juice "regardless of the fact that the juice must necessarily be watered to obtain flavor." Trumm did not contradict Brennan's assertion that essence is 999,600 parts water, but reminded him: "When you put cut-back in concentrate you are putting an awful lot of water in that." During Shaw's redirect examination of his witness, Trumm declared that the use of essence in pasteurized juice did not increase juice volume, lessen the quantity of juice solids, or generate any economic advantage other than greater quality and consumer acceptance.[31] To Trumm, juice dilution via essence was a nonissue.

But there were still other issues to defend. One involved the origins of the orange flavoring materials reintroduced into juice. The FDA was unsure whether these materials should be recovered from the same oranges juiced. Brennan asked Minute Maid's Brokaw: "Do you suggest that the essence added come from the oranges then being processed or the juice then being processed or do you include essences which might have been recovered from other juices?" Brokaw opted for the choice that was most inclusive: "I think it should be permissible to add essence from other juices in the same manner in which juices might be blended."[32]

If the FDA deemed flavoring material from outside oranges acceptable, the next question to resolve was how to classify the material. Brennan asked Beacham, "If one orange juice product or constituent of orange juice which was foreign to the orange juice which was being processed did not come out of the same batch of oranges from which this juice came, if this orange juice product were added to that juice what would this added orange juice product be?" Beacham responded that orange juice products such as orange essence and peel oil that originated from oranges "foreign" to those processed should not be considered "natural constituents." The proper designation for these substances was "optional ingredients."[33]

A final question critical to the consumer was whether orange-derived flavors should have to be listed on the juice label. Beacham gave the FDA's position that the addition of orange oil and essence "need not be declared on the label where properly used."[34] Later in the hearings he was more explicit: "Valencia orange oil obtained from California Valencias and added to Florida orange juice could be considered extraneous material. On the other hand, I have not suggested that label disclosure be made of that fact." This position was consistent, he said, with the "general philosophy which I have explained previously, and that is while it might be desirable to disclose every fact about the particular food product, as a matter of practicality it just can't be done . . . and so the Secretary has to make a choice as to optional ingredients to name." Even though he considered orange essence and oil "optional ingredients" rather than "natural constituents" of orange juice, he still did not think they warranted listing. The Secretary, he said, "must weigh the effect that such an addition produces" or the "significance of that effect and maybe other factors." For Beacham himself, "the addition of orange oil is not, in and of itself, a very significant fact."[35]

Trumm had a different reason for keeping essence and oil

off the label, stating that he didn't think "the consumer would have interest in knowing the method by which lost flavor and aroma has been restored to concentrated orange juice." Although Beacham did not challenge Trumm directly, he clearly did not agree. Beacham acknowledged that what to label was a matter of opinion and that "others might have a different opinion." Referring specifically to flavor addition, however, he was certain "that we could find some consumers who would desire that information."[36]

Maybe the FDA should have listened more closely to the consumers who desired "that information." Townsend testified at the hearings on behalf of fourteen million U.S. families against flavor use. If the FDA had given more weight to Townsend's words it might have retained more of a handle on a product that, as the following chapters will show, has fallen under the control of the big processors.

In the 1950s Kraft Foods Company, the National Association of Frozen Food Packers, and the FDA's Commissioner of Food and Drugs proposed definitions and standards of identity for orange juice and selected orange juice products. Between 1956 and 1957 the Federal Register issued notices of proposed rule making, setting forth the proposals. In 1960 the FDA published an order promulgating standards of identity for orange juice and nine orange juice products, including pasteurized, concentrated, and reconstituted orange juice.[37] Upon receiving numerous objections to the order, the FDA decided to hold hearings. The initial order was stayed, and over the first half of 1961 interested parties, mostly the FDA and orange juice processors, met to determine how to define standards of identity for orange juice and its processed imitations.

In 1963 the FDA published thirty-six findings of fact

from the hearings. First, the summary clarified that "orange juice" was the name to be used for unheated, freshly squeezed orange juice. It also identified a new product that was sold as "orange juice" or "fresh orange juice" and called "chilled orange juice" by the industry. It noted that the adjectives implied the heat-treated juice was a freshly squeezed product. The FDA preferred the adjective "pasteurized" to describe the orange juice. Finding number seventeen listed two justifications for federal regulation of orange juice products. One was the adulteration of orange juice products with water and sugar. The other was "the misrepresentation of reconstituted orange juice and of pasteurized orange juice as fresh orange juice." Evidence from the hearings was cited to support the conclusion that "the use of the term 'fresh' on commercially packaged orange juice or orange juice products would tend to confuse and mislead consumers."

The thirty-six findings were the preamble to a final FDA order establishing definitions and standards of identity for "orange juice" and nine orange juice products: "pasteurized orange juice, canned orange juice, sweetened pasteurized orange juice, canned sweetened orange juice, concentrated orange juice, sweetened concentrated orange juice, reconstituted orange juice, sweetened reconstituted orange juice, and industrial orange juice with added chemical preservatives."[38]

Part three of this book exposes the many failures of orange juice standardization. Fundamentally, standardization has not prevented orange juice processors, especially those involved in the production of pasteurized orange juice, from presenting their products as fresh.

The late Professor Thomas Mack with the author's dog, Dixi.

The father of Florida's orange juice industry, Edwin Moore was one of a team of three that invented frozen concentrated orange juice.

Although nursery owner Roland Dilley takes credit for inventing the container method for growing citrus rootstocks, where seeds are planted in a container and grown under controlled conditions, he still grows some of his rootstocks "bare root," as shown here.

Abraham, caretaker of Phillip Ruck's nursery in Frostproof, Florida, indicates the scar left by grafting, the thickening of the wood that has occurred where the incision was made in this tree.

Grower Jim Brewer samples a Valencia orange from his grove.

It is easy to see why Florida, and not California with its desert-like environment, became home to the juice orange and industry.

The history and current predicament of the mostly migrant
labor population (predominantly from Mexico) that picks
Florida's oranges deserve a separate volume.

To date, machine harvesters have been successful only with early season varieties such as the Hamlin. Until they can also be used for late season varieties such as the Valencia, the trees of which hold green and mature fruit concurrently, they will remain experimental, unable to compete with the discerning eye of the human harvester.

A Tropicana representative explains the orange picking process to Canadian retailers during an orange grove tour. According to Abraham (whose photo appears elsewhere in this section), a former orange picker who is now caretaker at an orange nursery, a worker can pick enough oranges in the course of a twelve-hour shift to fill approximately eight of these bins, at a pay rate of seven dollars per bin.

Orie Lee stands by a truckload of oranges weighing approximately
forty-four thousand pounds.

"Development opportunities" abound where citrus groves once grew.

III
Florida's Orange Juice
Industry Post-1960

X

Processed Orange Juice Hits Florida

I n 1961 the FDA acted on marketing tactics that misled the orange juice consumer. Jim Griffiths, an entomologist who has been involved in the orange juice business in Florida for over half a century, points to another problem with the hype that resulted from the invention of FCOJ: its impact on agriculture. "Everybody is trying to influence the consumer," he says, and such efforts are "probably successful to a certain extent. Basically orange juice demand has expanded and has had spurts that stimulated planting. And when [orange juice] gets overproduced it gets cheap and people drink more of it. And so that stimulates, the price goes up, and that stimulates planting."[1]

The cycle parallels a generic one described by agricultural historian Steven Stoll: advertising creates demand, which causes an increase in production, which leads to new and more urgent campaigns.[2] While Griffiths refers to an undefined "everybody" influencing the consumer, Stoll distinguishes advertising as the "body" building demand. True, processed orange juice tasted better than ever after FCOJ's invention. But inventive advertising was behind the consumer's demand for more.

Those like Edwin Moore, who expected that the invention of a palatable processed orange juice would free the grower from the bind of overproduction and consumers from the grind of juicing oranges, did not adequately account for the influence that huge increases in demand would have on agriculture. Griffiths, who as managing director of Citrus Grower Associates regularly deals with citrus growers, knew better. He has witnessed firsthand a weakness among farmers that makes exponential rises in demand dangerous: "Farmers don't have a good sense about managing supply. . . . [They] get too greedy. Like everybody else, the price goes up and they want more." In addition, he says, supervising supply has "never been done well by anybody."[3]

Jim Brewer, a grove owner and loyal supplier of Tropicana, summarizes the grower's dilemma: "We have to produce high numbers, whether we want to or not. We have to produce them or we can't stay. And that's our death too." At the same time he acknowledges that producing "high numbers" is not necessarily the better way. "That may be one of the answers that you come to in keeping Florida in the industry. . . . We may have to move from all these oranges down to a very nucleus of, a very high quality product."[4] His actions speak differently, however. He, like his neighbors, keeps planting more trees.

The orange tree's slow maturation complicates management matters by preventing growers from being able to respond immediately to rapid swings in juice demand. Maintaining supply and demand in equilibrium has thus remained mostly an elusive goal, especially for growers of juice oranges. Griffiths recalls one remedy to oversupply that was reliable for a short time. The years 1957 and 1962 are written in history as a period when "freezes pretty well took care of surpluses. We . . . depended on something happen[ing]."

Speaking positively about an event as damaging to citrus as a freeze sounds like heresy. But Griffiths is not the only one; Robert Behr, a senior economist for the juice cooperative Florida's Natural Growers, credits the natural disasters for the existence of Florida's orange juice industry. "Freezes," he says, "have come in to save industry." No amount of government or industry innovation has matched the effectiveness of weather's unpredictability. Now that irrigation technology has enabled growers to avoid the cold by moving their groves farther south, Griffiths doubts "that [will] happen anymore."[5] If the hurricanes of recent years are a sign of what is to come, perhaps fierce winds will replace icy conditions as insurance against out-of-control production.

Despite the industry's high hopes, statistics show that a good-tasting processed orange juice has not freed Florida orange growers from the self-destructive pattern of increased plantings followed by decreased prices. In the 2004–2005 season in Florida's southwest growing region, the cost per acre of growing a grove of Hamlin oranges for the processed market was estimated at $881.03. The gross revenue that each acre would bring in was estimated at $1,149.12, leaving a net return of $268.09 per acre. In central Florida the numbers were better: the net return was estimated at $529.81 per acre for a Valencia orange grove.[6] Nevertheless, after factoring in other fixed costs for expenses such as taxes and crop insurance, even the higher figure does not leave much left for the grower— including big operations owning thousands of acres.

Nurseryman Roland Dilley, who participated in Florida's transition from a fresh to a juice orange growing state, cautions that the state's citrus industry "is in a lot of trouble . . . certainly the worst I have seen in my years."[7] Dilley puts much of the blame on the Brazilian orange juice processors who are moving into Florida. The extent to which this group can be held accountable for the state of Florida's orange growers is debatable, yet their transformation of the orange juice processing sector is undeniable.

Florida led global orange juice production until the 1980s, when freezing temperatures intervened. Between the 1979–80 and 1981–82 seasons orange production fell from 206.7 to 125.8 million 90-pound boxes. Three years later another bout of cold weather reduced production by almost 50 percent of the pre-freeze figure, to 103.9 million boxes.[8] The state, not wanting to lose its increasing orange juice clientele, turned to Brazil, second to Florida in the production of juice oranges, to make up for the shortfall. It funneled money and men into the country to lay the infrastructure for a strong orange juice processing sector. Two men, Charlie Hendrix and James Redd, pioneered Florida's effort to build Brazil's orange growing and processing capacity. The two first traveled to São Paulo in 1962, when another historic freeze forced Florida to consider alternatives. The following year construction began on Suconasa, Brazil's first modern citrus processing plant.

Hendrix, previously a USDA chemist, trained the staff of the quality control laboratory that Redd set up for the Brazilian plant. A few years later the plant was sold and renamed Cutrale. Cutrale and soon-to-be-rival Citrusuco began to make their mark on the orange juice industry as Florida fell under the deep freezes of the 1980s. In the 1980–81 season Brazil was barely behind Florida in orange production, delivering 170 million boxes of oranges to Florida's 172.4 million boxes. By the middle of the decade Brazil had shot ahead, producing 239 million boxes, compared to the 119.2 million boxes that Florida salvaged from its freeze-desiccated crop. Brazil's lead steadily increased, climbing to a record 420 million boxes in the 1997–98 season, making Florida's record 244 million boxes for the same season comparatively unimpressive.[9] Griffiths suspects that if the Florida freezes of 1962 and the 1980s had not occurred, Brazil "might not have found out what kind of gold mine they had."[10]

As of 2004 Brazil's larger groves were producing oranges at about one dollar per box, while the cost was almost half again as much in Florida groves. A tariff on Brazilian orange juice imports currently protects Florida from a Brazilian takeover of the North American processed orange juice market. Citrus Mutual, a voluntary trade association of Florida citrus growers that organized in 1948 to maximize grower returns, has been vigorously campaigning to keep the tariff alive in a political environment that favors free trade. In 2004 the group spent four million dollars on such efforts, which were temporarily rewarded when, shortly before Election Day, George Bush landed in the hurricane-wrecked Citrus State to announce a commitment to conservative-leaning Florida citrus growers to retain the federal tariff on imported Brazilian orange juice in the face of impending World Trade Organization negotiations.[11]

Many, including a director at Tropicana (who, to preserve his anonymity, will henceforth be referred to as "Tropicana director"), expect the tariff to disappear eventually.[12] In the meantime, Cutrale and Citrusuco, still Brazil's major orange juice processors, have established ties with two of the biggest orange juice brands, Minute Maid and Tropicana. (Although these companies have operations in Florida, their head offices are elsewhere. Tropicana is a subsidiary of PepsiCo Inc., headquartered in New York. Coca-Cola Co., with its head office in Atlanta, has owned Minute Maid for almost half a century. The passage of Minute Maid, with roots in Florida Foods, the company connected to FCOJ's invention, into the hands of multinational corporation Coca-Cola Co. in 1960 is symbolic of the Florida orange juice industry's gradual makeover by foreign currency.) In the mid-1990s Coca-Cola began selling Minute Maid's processing plants to Cutrale. Allen Morris, a former financial consultant for Coca-Cola, was hired to perform an accounting of Minute Maid. Coca-Cola wanted guidance on how to cut the high cost of operating Minute Maid's juice plants. Morris found that processing plants have a 7–8 percent return on investment. "Why in the world would Coke put money in something doing 8 percent when it can put money in something doing 25 percent?" he said he asked himself at the time. The "something" doing 25 percent was the Minute Maid brand label. The solution for Coke, in Morris's opinion, was to outsource processing to a company for whom an 8 percent return would be worthwhile. Cutrale, which already supplied Minute Maid with Brazilian-processed orange juice, was an obvious choice. Lacking a brand label and the corresponding goodwill that makes the label profitable, Cutrale had good reason to accept the offer. Owning plants in Florida would give Cutrale not only access to the North American orange juice market but also a say over the future of Florida's seminal orange juice industry.

Thomas Spreen, an agricultural economist at the University of Florida, studies the effects that the infusion of foreign capital is having on the original structure of Florida's industry. Traditionally the industry was made up of smaller, decentralized growers and processors. The number of processing plants in Florida peaked in 1977 at fifty-three, and just over fifteen years later the number had dropped almost in half, to twenty-seven plants. Eighteen currently remain. Eight are operated by four corporations that are heavily invested in Brazil: Cargill, Cutrale, LouisDreyfus, and Citrusuco. Spreen estimates that these four corporations handle about 50 percent of the oranges crushed in Florida. The foreign consolidators have, Spreen concludes, upended Florida's citrus industry.[13]

Spreen singles out foreign investment and consolidation as the major forces transforming Florida's orange juice industry. But behind these forces is an equally important, complex web of small-scale interactions and relationships. Morris himself is one such example. His biography demonstrates his commitment to Florida's citrus industry for more than twenty years, performing services ranging from teaching Florida growers how to trade on the future's market to presenting lectures and publishing newsletters on understanding and managing the market for Florida citrus. Yet as an active advocate for the interests of the Florida grower and processor he shows no signs of remorse for encouraging Coca-Cola to make Minute Maid into a less local, more foreign-dependent, multinational corporation. In fact, he boasts about being the one behind Coca-Cola's sponsorship of a Brazilian takeover of Florida's citrus processing sector. Morris takes credit for Coca-Cola's sale of its processing plants to Cutrale, calling the move *his* "brainchild."[14] His double-dealing is indicative of the difficult straits in which the individuals involved in the transition of a local industry into a global entity are placed.

The Minute Maid–Cutrale partnership consummated, Tropicana and the smaller cooperative Florida's Natural became the Florida citrus industry's last bastions against Brazilian domination. Tropicana's "not from concentrate" ("NFC") has traditionally meant not from "Cutrale," or "Citrusuco," or from

any other Brazilian-owned company. Ana, the company's tropi-
cal fruit–bearing mascot, was still squeezing most of the oranges
in her basket well after Minute Maid had handed over this task.
Morris assumes that Tropicana continued to juice its oranges
because in the past it could not trust anyone else to make its
pasteurized juice.[15] Whereas Tropicana was one among a few
that pasteurized juice, many have long been adept at concentrating, the processing method that Minute Maid uses.

Now that NFC, the only category of processed orange
juice that was still growing at the turn of this century, has become the object of the juice consumer's desire, every brand and
processor wants a slice of the pie traditionally lorded over by
Tropicana, along with Florida's Natural.[16] Coca-Cola's "Simply
Orange" has taken a big piece. Even Brazilian-based processors,
who in the past produced only easy-to-export concentrate, are
taking a share of NFC. These processors have long recognized
Florida's need for NFC in the summer—when the state's oranges are fermenting on the tree—and fall—when Florida's oranges, mostly Hamlins, are pale when squeezed. With a growing
season that is Florida's opposite, Brazil has ample and better-
colored juice during Florida's summer and fall months. The obstacle to meeting Florida's NFC demand has been the difficulty
of finding a vehicle that could carry enough volume to make
shipping economical. But Brazilian funds have now sponsored
the construction of two vessels that have space for NFC travel.
The Tropicana director marvels at this technological feat:

> When I came to work in this industry, which
> was 1988, Brazil was shipping concentrate to North
> America in vessels that carried, I think, two million
> gallons of concentrate. And they were vessels that
> had been freighters that were converted; tanks were

> put in. And those things were incredible. A few
> years later they went into vessels that carried three
> million gallons. And they were designed to be tanker
> vessels. And those vessels were incredible. The lat-
> est thing that Citrusuco has is vessels that can . . .
> load in Santos aseptically, go to Europe, unload into
> tank farms aseptically, not pasteurize it again, and
> they're seven million gallons. And they're incredi-
> ble. What's there going to be five years from now?
> What we've seen so far is that these vessels become
> economically out of date before they're engineering
> out of date.[17]

With vessel storage capacity having reached the seven-million-
gallon mark—the equivalent of the total juice Tropicana could
store on land when it began using tanks in the 1980s—the ex-
port infrastructure was in place for Brazilian processors to
pasteurize juice. Advances in shipping technology explain the
increase in Brazilian NFC imports from zero at the end of
the 1993–94 season to 65,007,378 gallons, valued at $68,766,367,
at the end of 2007.[18] Now that it is moving overseas not
even NFC, once a Florida pedigree, is a purely Florida product
anymore.

If shipping technology continues to progress so rapidly,
Tropicana will no doubt step down from the processing sector.
Like Coca-Cola has done with Minute Maid, PepsiCo., which
bought Tropicana in 1998, will turn Tropicana into a purely
juice-branding venture. It will let another company worry about
the low investment returns and uncontrollable weather that
can make the processing of an agricultural product an un-
glamorous pursuit. Tropicana, which is already purchasing
Brazilian NFC, appears well positioned for a Coca-Cola-style

stripping down. In February 2005 Florida news wires reported on talks between Citrovita Agro Industrial Ltd., Brazil's third-largest juice processor, and Tropicana. Under the terms of the reported contract under negotiation, Tropicana would provide a construction loan to Citrovita to build a plant in Florida and commit to purchasing the Brazilian company's juice. Tropicana spokesman Peter Brace denied the reports.[19]

Declining to confirm or deny the rumors, the Tropicana director looks to Minute Maid's sale of its plants to Cutrale as a model for Tropicana: "I think they [sold to Cutrale] very clearly to become a marketing company and to get the heavy manufacturing and all of the low return investment off their books. . . . They need to be making 20 percent a year, [they're] a privately held company. . . . In today's day and age you can make 8 to 10 percent a year and be very happy with it." But companies such as Coke and Pepsi, which "have to report to Wall Street every ninety days, can't do it." He suspects that this financial reality is why Coca-Cola "got that off the balance sheet," and he said that "there are some suggestions that Pepsi might do the same thing." Citrus nursery owner Roland Dilley does not want Tropicana to hand juice processing over to Brazil. "If they do that, well, you can just fold your tent here in Florida," he said, adding that his nursery would likely be dug up for an industrial park, a move that would literally uproot the next generation of Florida Sunshine trees. Edward Smoak, a second-generation grower, agrees: "I'd say this is a darker cloud than we've ever been under."[20]

Regardless of what Tropicana decides to do with its processing plants, the Tropicana director is certain that if Brazil builds more ships and exports more NFC to the United States, "We'd be a part of it." At the same time, he acknowledges that such a development would be a "threat to the Florida industry."[21] Although Florida's Natural NFC citrus juice cartons are still stamped with the state-of-origin seal of the Florida Sunshine Tree, the cooperative's reconstituted juice cannot sport the 100 percent Florida label anymore; it contains Brazilian imported concentrate. To stay competitive with the most profitable NFC producers, the cooperative of Florida citrus growers and packers may find itself following leaders, such as Tropicana, that are importing Brazilian NFC as well as concentrate.

Brazil's influence over Florida's orange juice industry extends beyond processing. Its bounteous supply of land for growing juice oranges is diminishing the value of Florida's groves. Once gold mines, Florida's orange groves are now undergoing a reverse alchemy as they are turned into firewood to make room for the construction of more lucrative condominiums. One irreplaceable citrus resource under the supervision of Orie Lee stands precariously. An orange grower and researcher (and self-professed "old country boy"), Lee, in partnership with University of Florida and USDA citrus breeders, has traveled the citrus regions of the world since 1970 in search of insect, disease, and weather-tolerant citrus varieties. The most valuable finds have made their way into Lee's experimental citrus variety collection, which, he estimates, includes as many as eighty different types of processing oranges. A few of these varieties have attracted the attention of private orange juice processors, and their effect on recent planting trends is noticeable. One selection, the "Midsweet," released from the collection in 1987, became the third-most planted variety in the state. In addition, he is proud to say that almost all of Florida's approximately 224,000 Vernia orange trees are the progeny of three or four in his grove.[22]

He is less enthusiastic when he talks about the market value of the land on which his collection sits: "This land is worth today three times as much for houses per acre as it is for raising oranges." Once he retires, he says, "It'll be gone," as he notes that his children are pursuing careers unrelated to the orange-growing industry. Jude Grosser, one of the University of Florida citrus breeders with whom Lee has worked, recognizes that growers like Lee "are an endangered species."[23]

With more and more growers selling out to real estate developers, Florida citrus is losing its link to the state's identity. Andy LaVigne, CEO of the grower association Citrus Mutual, compares Florida to other agricultural states: "You drive through the Midwest, you see amber waves of grain. Vacation in northern California, you see miles and miles of vineyards. Here, you used to see oranges. But you no longer see them. You really have to get off the beaten path to see production here, and that turns out to be a problem."[24] Nowhere is this more apparent than on the drive to Lee's grove. From Highway 92 take 17 South, which is the same exit, a billboard announces, for a Nike outlet store. This is St. Cloud, a suburb of Orlando, home to Disney World, Florida's tourist heartland. Several turns off the main road, tucked well out of sight behind an endless strip of flashy attractions, is Lee's grove, which his parents planted sometime around 1910, and which now runs the length of Lillian Lee Road (named after an ancestor). It remains unseen by the hoards of people driving along Highway 17, who can see only the concrete and glitz that are the land of Disney.[25]

In the 1950s Lee's county produced forty percent of Florida's oranges. Since then, most growers have retreated to the state's southern interior and the county now grows only six percent of the state's oranges.[26] Malls have replaced the orange juice stalls that once lined Florida's thruways. With no groves in sight, tourists, along with Florida residents, are unlikely to care about the grower's plight.

As the distance between citrus grower and urban dweller has diminished in Florida, the new neighbors are more at odds with one another than ever. The Florida citrus grower and non-grove-owning landholder have a lot to gain by working together, but planter and transplant are having trouble seeing past their differences. Residents complain about the state's intrusion into their gardens to cut down citrus trees infected with citrus canker, a virus that harms only the commercial grower. Growers react to these mostly recent Florida transplants with bitter words about urban land use. Jim Brewer, orange grower *and* cattle farmer, tells the story of how he sold his neighbor some of his pasture, which she covered in concrete and so the land no longer can collect water. At an environmental meeting that the two subsequently attended, the woman remonstrated *him* for wasting scarce water on his grove. Farmers, he concluded, "are mostly minorities, so they don't have a voice."[27]

Sharon Garrett expresses the sentiments of many small growers who feel they've lost their voice. She is especially outspoken about the change Tropicana is making in how it pays its growers. The company used to pay growers for their oranges based on the pounds of orange sugar solids that could be extracted. Now Tropicana prefers to base payment on the number of gallons that can be squeezed out. Garrett says the shift "will take some of the rest of us out of business. I think they've figured out another way to steal more from you."[28]

Another small grower, Rory Martin, who runs his family's twenty-five-acre citrus grove and who had a contract with Tropicana, shares Garrett's view. Noting that the percent of sugar solids contained in a box of citrus scarcely fluctuates compared to the number of gallons, he says that Tropicana is trying to "lock down the price per load." And because he grows a high-quality orange, one with a lot of sugar solids, he will lose out. If Martin wants to continue being paid on the basis of pounds solids, the best he can get is a three-year contract at eighty-five cents per pound solids, with the added condition that he must sell exclusively to Tropicana. The offer, he says, gives him no choice but to opt out.

Martin said that the relationship between growers and the company has changed radically. His grandfather signed on with the late Anthony Rossi, Tropicana's founder, and had only praise for the legendary figure, whom he described to Martin as a "man of uncommon character." The company used to support the community, Martin said. He remembers when grower and processor socialized together: "Once a year [Tropicana would have] growers over [for a] barbecue lunch." He smiles nostalgically: "Tropicana was that kind of company."[29] Allen Morris, who worked for Tropicana after the Rossi years, admires the founder's progressive approach toward his growers. When orange prices dropped below the cost of production in the 1970s, Rossi, who still owned the company, presented a five-year fruit contract that offered growers facing huge debts almost double the debilitating market price. "He had the foresight to understand that in reality processors' treatment of their grower suppliers should be no different than if that were their grove division and they owned it, because in essence they're intricately linked with each other," Morris says.[30]

Rossi's priorities were the opposite of today's leadership. For example, Morris quotes Rossi's approach to advertising: "If it's a good product the consumer will buy it. I don't like to brag about it."[31] But once Pepsi took over and began moving big dollars into marketing, the family Rossi fathered disintegrated. The annual barbecue became an event to remember.

Symbolic of the Florida grower's declining status, on April 20, 2005, eleven of the twelve members of the FDOC's governing body, the Citrus Commission, voted to change the agency's 1993 mission statement, which charged the FDOC with "enhanc[ing] the welfare of Florida's citrus growers and the groves they operate."[32] The new statement codifies the broader constituency that the FDOC believes it represents. It says the FDOC's purpose is to: "Grow the market for the Florida citrus industry, to enhance the economic well being of the Florida citrus grower, citrus industry and the state of Florida." At one time citrus growers *were* the citrus industry. Now the citrus industry is a separate entity.

In the talks leading to the revision, growers and their sympathizers clung to the idea that the FDOC exists mostly for the grower. They argued that because growers pay the taxes that make up most of the FDOC's annual budget, they should be the focus of the FDOC's concern. Pat Carlton, Wauchula grower and sole naysayer in the Citrus Commission's vote, was the only grower representative on the commission. He thought the statement needed to give the grower greater weight.

While growers waved their tax dollars, the commission majority said the importance of orange juice marketing, which absorbs most of the grower's money, places the processor—and the professional advertiser—with substantial influence over the citrus industry, including its growers.[33] The implicit conclusion, that the processor should receive a proportional amount of the FDOC's attention, suggests that the "citrus industry" mentioned in the new mission statement is not only separate from the grower but a pseudonym for the processor.

If the commission looked like it did in 1935, its membership consisting largely of growers, Carlton would have been in the majority. However, the grower is no longer the commission's

Florida's Orange Juice Industry Post-1960

anchor. FDOC informational material explains that "some [on the commission] are growers, some are fresh fruit packers, and some are from juice companies."[34] The outcome of the vote shows that "juice companies" have replaced the grower as the guiding force behind the FDOC.

Grower advocate Jim Griffiths, an industry icon, says that the words "Florida citrus industry" "do not mean the same as sixty-five or fifty years ago." In an article for Lakeland's *The Ledger*, he is quoted as saying: "Then, it was a Florida citrus industry owned by Floridians. That's not true today at all." According to the article, Brazilian-based companies such as Cutrale Citrus Juices U.S.A. Ltd. and Citrusuco North America Inc. hold almost 40 percent of Florida's citrus processing capacity.[35] The "citrus industry" that the FDOC has welcomed to the bargaining table is a Florida-outfitted foreigner. By writing "industry" into its mandate, the FDOC has thus effectively given Brazilian processors a state-sanctioned, permanent foothold in Florida.

During the mission statement debates Griffiths doggedly defended the interests of the Florida growers, whom, as managing director of Citrus Grower Associates, he represents. For one thing, he complained that the opening line, which charges the FDOC to "grow the market for the Florida citrus industry," is ambiguous and asked FDOC Citrus Commission Chairman Andy Taylor what "market" the FDOC was responsible for growing. Taylor acknowledged that the wording was nonspecific and answered that "market" meant any market selling Florida citrus products. "Florida citrus products" can be defined multiple ways. At one extreme, any citrus product moving from Florida could be considered a Florida citrus product—including products processed and packaged by foreigners in Florida, such as products that are mixtures of Florida and

Brazilian juice, as is the case for most reconstitutes—provided it is packaged in Florida. At the other extreme, only products made and owned by Floridians and derived from 100 percent Florida citrus could be called a "Florida citrus product." The middle ground would be that a product not made or owned by natives but containing 100 percent Florida citrus is a "Florida citrus product." With so many possible interpretations, Taylor's intended clarification only generated more uncertainty surrounding the FDOC's decree.

Griffiths countered that Taylor's response was defining the "market" as any market "selling the products processors are selling." Interpreting "market" so broadly could justify the state spending Florida grower tax dollars on building the market for foreign-based juice processors and their suppliers.[36]

The divinations of those like Dilley, Lee, and Griffiths for Florida's citrus industry are overshadowing the initial appraisal of FCOJ as a Cinderella product. Since FCOJ's coming of age, processed orange juice has grown at the expense of fresh orange consumption. In 1947, just before FCOJ's debut on a major scale, per capita consumption of fresh oranges was 39.4 pounds. In the same year per capita consumption of processed orange juice was the single strength, or reconstituted equivalent, of only 0.48 gallons. At the close of 2004, per capita consumption of fresh oranges had fallen to 11.19 pounds, while per capita consumption of orange juice had grown to 4.77 gallons.[37] The transformation from a fresh-citrus-eating to a processed-orange-juice-drinking nation was supposed to save Florida's citrus growers. However, if growers were fighting for their livelihoods prior to FCOJ's midnight arrival, they are in no better shape decades later.

In January 2006 the Citrus Commission decided to form a blue-ribbon panel to advise the commission on the future of

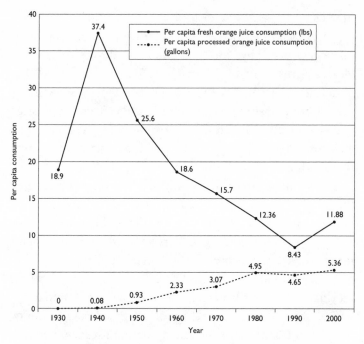

American consumption of fresh oranges dropped dramatically
when frozen concentrated orange juice was introduced in the 1940s.

Florida citrus. In an article in the *Lakeland Ledger* about the
FDOC action, Robert Underbrink, executive officer of one of
the state's largest citrus groves, cited the historical importance
of citrus to Florida's economy, but he also warned: "It won't
be a nine billion [dollar] industry, I can tell you that. . . . The
time is right. You better get a committee together. It's going to
be different. It's going to be different real soon." He said that
disease, citrus canker in particular, is diminishing fruit production while commercial and residential real estate developers are making tempting offers to Florida's grove owners.[38] He

did not talk about the encroaching influence of the Brazilian processor, another factor in assessing the staying power of Florida citrus. He also did not mention that the invention of a heavily consumed processed orange juice is not serving the interest of the Florida grower.

Florida growers are not the only victims of the success of FCOJ and its processed orange juice offshoots. The following chapters explore how advances in processing technology, combined with smart advertising, are subverting FDA orange juice regulation to the detriment of the "interest of consumers."

XI

NFC Orange Juice
Pours into the Nation

The FDA disregarded processors' protests over whether to use the word "pasteurized" to describe pasteurized orange juice. Section 146.140 of the Code of Federal Regulations introduces the product's standard of identity with the title "Pasteurized orange juice," and the body of the text reiterates this: "The name of the food is 'Pasteurized orange juice.'" It stipulates that the word "pasteurized" must be "shown on labels in letters not less than one-half of the height of the letters in the words 'orange juice.'"[1] Clearly the intent was to make "pasteurized" part of the public's orange juice vocabulary. The regulation communicates the FDA's belief that labels carrying more production-related information would help end orange juice marketers' pretense that heat-treated and hand-squeezed juice are essentially the same thing.

Orange juice consumption statistics reveal that the appearance of the word "pasteurized" on orange juice cartons has not, as processors feared, diminished the product's marketability. In fact, pasteurized orange juice has been the single standardized orange juice product among the family of twelve that has recently been gaining market share.[2] On cartons of Tropicana Pure Premium the brand "Tropicana" is scribbled in large type and trademark style across the top. "Pure Premium" floats on a ribbon beneath. In smaller, italicized letters, "Not from Concentrate" completes the logo. Section 146.140 notwithstanding, this last phrase, "Not from Concentrate," which Tropicana coined, has supplanted "Pasteurized orange juice" as the product's common name. On the carton's bottom corner, in small type, the word "Pasteurized," which was the cause of so much controversy during the 1961 hearings, appears unobtrusively. Its inconspicuous positioning seems to contradict the spirit of the regulation, which requires that the word "pasteurized" be "shown on labels in letters not less than one-half of the height of the letters in the words 'orange juice.'" Presumably anticipating that the industry would want to reduce the impact of the word, the FDA set a minimum stature, using the seemingly essential product descriptor "orange juice" as the referent point for the size of the word "pasteurize." Evidently it did not anticipate that the words "orange juice" would lose their significance as a product descriptor. However, on Tropicana's carton the words "orange juice" are written in italic type and in relatively faint script on the very bottom of the label, and the word "pasteurized" is located even more inconspicuously, detached from its referent "orange juice." The regulation may even explain Coca Cola's name for its not-from-concentrate: "Simply Orange." Adding "Juice" would require substantial changes to the label, as "pasteurized" would then have to be more noticeable, printed in type that is at least half the size of the brand slogan. By deemphasizing—in this case by deconstructing the term "orange juice"—orange juice manufacturers have rendered impotent the problem word "pasteurized."

Industry consultant Allen Morris, a one-time Tropicana

employee, narrates an encounter with an orange juice buyer that confirms the processor's success in effectively neutralizing this regulation. During an elevator ride the head of the litigation department of a major financial house struck up a conversation with him. The woman, in her late thirties, asked, "Didn't you used to work for Tropicana?" After he replied that he did, she told him, "I love that Tropicana orange juice, I only serve my family fresh squeezed juice." He was struck by her comment. "Here's a woman with a law degree and MBA from Harvard, [who] earned her living trying to pick apart facts as a litigator, and she thought Tropicana Orange Juice was fresh squeezed. What do you think the average housewife thought?"[3]

On paper, the standard of identity for pasteurized orange juice attests to FDA resistance to the processor's demands for less demanding labeling standards. But despite the FDA's implementation of standards that were designed to reform, processors managed to maintain naming practices that adhered to industry norms.

In one respect, the processors were insightful when they insisted during the 1961 hearings that confusion would reign if the FDA changed pasteurized orange juice's name. The processors recognized the consumer's heavy reliance on marketing in differentiating between good and bad juice. The history of the phrase "not from concentrate" proves the impact that a few words can have on the decisions consumers make.

On its own, "not from concentrate" contains none of the melody that makes marketing jargon so memorable. The phrase is prosaic, even unwieldy. It was not the result of the type of systematic study of consumer psychology that generally precedes any change to a processed food's name. Instead it was the unexpected brainchild of an off-site meeting among a select group of Tropicana employees in July 1985.

The topic under discussion was Tropicana's pasteurized orange juice and its difficulty in keeping pace with sales of reconstituted orange juice, or "recon."[4] The rising popularity of recon was troubling to pasteurized orange juice processors like Tropicana. Historically Tropicana dominated the "Ready to Serve" (RTS) processed orange juice market. Pasteurized orange juice was the major contender and Tropicana the biggest manufacturer of the single-strength, RTS juice. The primary competitor was frozen concentrated orange juice, a product for which convenience had originally been a primary selling point. As Robert C. Evans, then secretary-manager of the Florida Citrus Commission, stated at the annual meeting of the Florida State Horticultural Society in 1944, FCOJ's promise lay in its convenience. He predicted the juice would lead to "substantial increases in the percentage [of oranges] used for processing" because "the housewife is generally becoming more and more allergic to preparing foodstuffs, regardless of how little effort is involved."[5] Although consumers were still buying fresh oranges from Florida to make juice right up to the introduction of FCOJ, Evans concluded that they were willing to spend the time and energy only because the one alternative at the time was canned juice, which was infamous for its dreadful taste. He correctly anticipated that FCOJ, which retained some of the taste of fresh juice, would turn home squeezed into history.

Had Evans known that the invention of flash pasteurization would produce a ready-to-serve juice that tasted good, he might have guessed that FCOJ would lose its competitive edge. Director of Citrus Grower Associates and industry expert Jim Griffiths notes that, with flash pasteurization, the reduced heating time "retained the flavor better" and that the process is "less harmful to the juice."[6] FCOJ was economical but, compared to the no-longer-burnt-tasting pasteurized juice, a has-

sle to thaw and dilute. Sure enough, sales of pasteurized juice
began to overtake those of frozen concentrate, but frozen orange
juice concentrators had a solution: they decided to reconstitute their concentrate for the consumer. Recon thus began to
appear alongside pasteurized juice in the same dairylike containers in supermarket refrigerators.

Recon has a major advantage over pasteurized juice in
that it is less expensive to produce. Made from space-saving
frozen concentrate, it stores compactly. Processors usually add
water only at the point of distribution or, depending on infrastructure, retail. Pasteurized orange juice processors do not
have that option. To deliver juice every month of the year they
have to invest in structures large enough to hold their volumi-
nous single-strength juice. Allen Morris notes the substantial
cost differential between storing the concentrate from which
recon is made and NFC: "[The storage] of concentrate is probably a penny/pounds solid/year, and to store NFC is 20–25
cents/pounds solid/year."[7]

The high cost of storing pasteurized orange juice was
only part of the reason Tropicana Pure Premium was having
trouble competing with products such as Minute Maid's recon.
The freezes that hit Florida in the 1980s, striking in 1981, '83,
'85, '87, and '89, devastated the state's citrus industry. Concentrators began importing Brazilian concentrate to make up for
the shortfall. Tropicana, in contrast, had no backup for its
pasteurized product, as the technology did not yet exist for
shipping large vats of liquid juice. Tropicana therefore had
to rely solely on the scarce and expensive supply of Florida
oranges. To stay in business, it started to charge more for its
pasteurized product. According to the Tropicana director
quoted earlier, a story circulated through the company that
officials there "really began to differentiate their product either

when they had to raise their price or after. And then they began to build [NFC] as a point of difference." Market researcher AC Nielsen, the agency that has tracked orange juice sales and prices since the early 1970s, did not differentiate between "not from concentrate" and "reconstituted" RTS orange juice until 1990–91, so it is impossible to verify whether the introduction of the term "not from concentrate" coincided with the sharp rise in the price of pasteurized orange juice. Nevertheless, this Tropicana official is satisfied that what he has heard "makes for a pretty good story."[8]

Morris and the Tropicana official may disagree over whether Tropicana's primary motivation to differentiate its Pure Premium pasteurized orange juice was the price paid for storage or for the raw material. But they both agree that by 1985 Tropicana was feeling the heat. In July 1985 Tropicana's president, Spencer Vogue, gathered the company's officers and directors for a meeting that marked the beginning of pasteurized orange juice's rise to most popular orange juice in its class. Morris was among those behind Pure Premium's facelift. He and his boss, a chief operating officer, were on opposing sides of a debate over a report that a major marketing firm had prepared for Tropicana at a cost of more than $1 million. The report concluded that Tropicana should shift to manufacturing and marketing reconstituted orange juice only.

Morris himself recognized that a switch to recon might have been better from a quality standpoint. "Actually you can make a better recon than [pasteurized juice] if you do a good job," he says. "You can hide processing mistakes in an evaporator that you can't with [pasteurized juice] . . . if you get some bad fruit that you didn't detect in the grading process . . . you just can't afford to keep correcting for that and so in essence the recon will taste better."

Nevertheless, Morris was among those who were against taking the recon route. He argued that "the consumer wants something, the closest they can find to fresh-squeezed, and you can market . . . the concept of an orange juice that [hasn't] been manufactured as much and it will have some merit." He and his allies wanted to play up the "Pure" in "Pure Premium." By the end of the day Vogue was convinced. As Morris tells it, the president decided: "We're going to give this idea another year but we're going to do one thing differently, we're going to tell the consumer that our product is not made from concentrate." Persuaded that consumers judge their juice by the cover, he decided to tinker with words rather than with substance.

Vogue's decision not to buy into concentrate technology turned out to be wise. But the road to rewriting his company's product was divisive. Vogue was being literal when he said the company would tell the consumer Tropicana's product was "not made from concentrate." But according to Morris, the "big marketing people" were critical: they "hated 'not from concentrate' because they said it was 'not' something." Morris recalls that Vogue set up a "big contest" to determine a better word choice. But nobody could come up with a superior alternative. And that is how "not from concentrate," an illegitimate to the master marketer, became Tropicana's poster child.

Thanks to "not from concentrate," the market for Tropicana's Pure Premium juice exceeded everyone's expectations. Morris relates how the new "marketing guy" Tropicana hired to sell the refashioned product organized a campaign to tell the consumer "that Tropicana had a very special product called 'Not from Concentrate Pure Premium,' it has never been concentrated and manufactured as much as the other brands had," and that "Tropicana was the only source of it."

Over the next five years, these three words helped to double the company's sales volume and almost triple its profits. "It was incredible, the response was absolutely phenomenal," Morris says. Although AC Nielsen does not track the volume and sales of individual companies, it does show that the volume of "not from concentrate" sales jumped from 125.7 million gallons, valued at $652.9 million, in 1990, to 216.4 million gallons, valued at $1.03 billion, in 1995.[9] Morris quotes comparable percentages for Tropicana alone.

Taste tests confirmed the reason behind those impressive numbers. "One lady said, 'I don't care how bad it tastes, I don't care if it tastes awful, I want, and my customers want, NFC,'" Morris says. According to Morris, the groups could not differentiate NFC when sampling different kinds of unlabeled orange juice. "But if you label one carton NFC and the other FC [frozen concentrate] and make NFC taste worse they still pick the NFC." He remembers the tasters' conclusion: "It didn't taste as good but I know it's better for me and better for my family." He speculates that "it isn't the taste that makes NFC popular, it's the *concept* of being closer to fresh-squeezed that makes it more popular."

Griffiths has but one explanation for the category's sudden popularity: "It was perceived as being fresh-squeezed and that was better than having to go through some terrible process of concentrating and then adding back water. *Perceived* is the big word there." He is certain that the reason for NFC's success has been "primarily marketing."[10]

Indeed, Louise Lee, wife of orange grower Orie Lee, does not think that even she could tell the difference between recon and NFC. Although she and her husband live modestly, surrounded by their generations-old orange grove, Louise Lee does not hesitate to pay the substantial premium of $1.40 per gallon for NFC—though not for any particular reason: "I don't know, I just glommed onto this 'not from concentrate.'"[11]

Morris has no doubt about whether NFC tastes better than recon: "There ain't a difference."[12]

The prehistoric hunter and gatherer judged good from bad produce based on skin color. Supermarket-goers, having inherited their ancestors' reluctance to use taste as more than a second measure of quality, rely on product labels as accurate indicators. A careful consideration of looks and cautious trust of taste served a purpose for the gleaner, who had to be alert to the possibility of poisonous fodder. An advantage of the modern regulated food environment is that consumers can rest somewhat assured when eating store-bought food. Although unhealthy foods dressed to entice abound, potentially lethal ones have been kept to a minimum (yet if the recent spate of food scares is any indication, this statement may not continue to hold true). The reverse of what was beneficial to the forager is thus advantageous to the modern shopper. Because the processed food label is neither as predictable nor as honest as nature's container, taste is a safer, if not foolproof, gauge of goodness than is scrutiny of outer wrapper. Yet born with the scavenger's genes and brought up in an environment plastered with advertisements, the modern consumer's brain is trained to suppress the senses and allow the mind to take over. NFC's command of the processed orange juice market is filled with reminders that the reordering of the senses will not be easy for consumers who, hardwired to eat with their eyes, do not prioritize savoring the flavor.

XII

The Orange Juice Wars

Most know that Tropicana Pure Premium *is not* from concentrate. Few know what it *is*. Consumers of the Simply Orange brand of not-from-concentrate orange juice may have heard the company's motto: "Simply unfooled around with." Few question what the statement means, let alone its accuracy. Without a source of information other than the product label, yesterday's homemaker and today's orange juice buyer must accept the cartons on the market knowing very little about their contents. This is problematic because of the rift that exists between the reality of processed orange juice and retail rhetoric. Take NFC, for example.

Although NFC is the crown jewel of the processed orange juice empire, there is division within the kingdom. Simply Orange, progeny of Coca-Cola, is vying for the throne currently occupied by PepsiCo's Tropicana. The Tropicana director succinctly describes the relations between the two: "Pepsi and Coke hate each other."[1] Though the war may be between the parent companies, the animosity has filtered down to the subsidiaries. The consequence, which is the orange juice equivalent of the cola wars, has incited marketing campaigns that have persuasively presented orange juice fiction as fact.

First a bit of history. Minute Maid reconstituted orange juice was Coca-Cola's first in line. But when PepsiCo's wunderkind Tropicana "Not From Concentrate" began to steal the show, Coca-Cola fathered "Simply Orange," the company's second and more successful attempt at a not-from-concentrate orange juice. In April 2003 it opened a multi-million-dollar packaging plant for Simply Orange in Auburndale, Florida, former home of Minute Maid. In April 2005 it announced it would add another production line to accommodate Simply Orange's growing popularity, promising that the addition would create dozens of new jobs.[2]

The investment signaled a shift in Coca-Cola's priorities. Historically known for its secret recipes, the company began to refocus its attention from inner enrichment to outer appearance. It designed an award-winning carafe for Simply Orange that now comes in two sizes: the original 1.75 liter container and, in April 2005, a single-serve 400-milliliter clone. Ronald Muraro, an economist with Lake Alfred's Citrus Research and Education Center, is enthusiastic about the carton: "It's not the square carton anymore. . . . The milk industry came up with 'chugs'. . . . If you go look in the store, look for TG Lee milk, they're neat little boxes, and they welcome you to grab it and chug it." Remarking that "Coke did come up with a real attractive container for Simply Orange," he wonders "why the other juice people didn't pick up on that . . . [and] come up with something else which is clever and attractive."[3]

The Pepsi family responded to Coca-Cola's plans to enlarge Simply Orange's packaging plant by announcing that Tropicana would add fifty million gallons of juice storage capacity to its Bradenton, Florida, plant. The expansion would require building 3.2 additional acres of storage, for a total of 8.7 acres. With the city planning commission's cooperation, Tropicana would be able to store nearly double the existing sixty-million-gallon capacity. The five-member commission quickly and unanimously approved the request.[4]

This struggle for the NFC throne provides the background for this chapter about the duplicitous advertising that is fueling NFC's rule over all products orange juice. Simply Orange is an appropriate product with which to begin assessing the integrity of NFC advertising. The product is not simple. Neither is it necessarily only orange. During the fall Hamlin orange season, for instance, many NFC producers add tangerines to their "100 percent orange juice." Tangerines contribute desired hue to the otherwise pale Hamlin squeezed juice. The practice is legal, so long as the citrus hybrid does not exceed 10 percent of the total juice content.[5]

The fact that Simply Orange is literally not simply orange is symbolic of more serious misconceptions. Clicking on www.simplyorange.com in 2006 took you to the home page (which has since undergone some minor modifications) and drew your attention to the left side, where, in white letters contrasted against orange background, Simply Orange's trademark read: "Simply unfooled around with." With scant other text on the page, the slogan was especially noticeable. Although it has been erased from the 2007 Web site, the newer one remained fundamentally the same. An enlarged image of an orange, with stem and vicariously sweet-smelling blossom intact, appears inside a horizontal screen spanning the width of the Web page. Behind it the scene fades between an orange tree fat with glistening fruit and the uniquely crafted carafes of Simply Orange. At the screen's bottom right-hand corner the words "Gently Pasteurized," camouflaged in tiny and slightly more deeply colored orange letters than the yellowish orange background on which they appear, ask to be overlooked.

These words indicate that Simply Orange is not being completely straightforward when it says its juice is "unfooled" with. Those in the industry know what NFC processors must do to provide a year-round, nationally available, sixty-day shelf-stable product. The almost sixfold increase in the shelf life of pasteurized juice, from ten to twelve days forty years ago to sixty days today, alone suggests there is nothing "simple" about the process. The juice has been pasteurized and not "gently," as advertised. But pasteurization is neither the only, nor necessarily most life-altering, step of NFC manufacture.

Before joining the Citrus Research and Education Center, where she regularly consults with processors in the industry, Renée Goodrich worked for Ocean Spray, and before that Tropicana, where she was a processing engineer involved in research and development. In a processing plant everything looks unfamiliar to the average consumer, including the seemingly straightforward practice of juice extraction. The steely brutality of the procedure cuts through Goodrich's explanation of when and why processors add flavor back. She says that at certain times the oranges in the mix could produce a juice with a perfect oil level and profile, but the manufacturers require that oil levels be perfect 100 percent of the time. "What we typically do though is extract fairly hard for yield," she says. "The juice ends up with a bit too much oil and we're allowed to either take it off by vacuum-stripping or also deoiling centrifuges. . . . When you do that you also take out more than you might ordinarily do just so you can then add back your desired oil flavored package." "Hard" extraction is not unique to any one company or category of juice. Processors of both pasteurized and concentrated juice do it, Goodrich says, for "economic reasons." High yields are especially important lately in what she describes as a "fairly low margin industry."[6]

The extractors Goodrich identifies allow a plant such as Tropicana's in Bradenton to process as many as 365 trailer loads of oranges per day.[7] The contents of each truck that grower Orie Lee fills for Coca-Cola can have "almost 4,000 pounds of sugar . . . about 550 boxes, about 44,000 pounds."[8] The numbers that Tropicana cites are similar. According to the Tropicana director, "A truck can carry from 500 to 650 boxes of fruit, depending on length, weight of cab, etc. A good average is probably around 550 boxes, which at 90 pounds each would make an average load weigh 49,500 pounds." At the level of the individual extractor, with 365 trailer loads per day this figure breaks down into 375 oranges squeezed per minute, for an average 2,500 liters of juice produced per hour, the rate typical of Tropicana's extractor.[9]

Film equipment or recorders are not allowed on most plant premises to document the details. Important patrons or, with some luck, an interested person may be able to arrange a tour, in which earplugs are needed to protect against the deafening noise that strikes from all sides once someone steps inside the concrete structure where hundreds of thousands of oranges are squeezed each day. Juice extraction was not always so dependent on heavy manufacturing equipment. Wallace Roy, who was in charge of technical services at Minute Maid in 1961, remembered former ways while testifying at the orange juice standard of identity hearings: "Juice at one time was produced by hand reaming. Women squeezed halves and were paid by the bucket of extracted juice."[10] The technology has advanced so far beyond the original manually operated reamer that the orange, once massaged by hand, is now roughly gripped by automatic steel cleavers.

If the reality of hard extraction is at odds with the descriptor "gently pasteurized," so is the manner by which processors ensure endless orange juice refills. Processors who want to supply sufficient juice to meet the nation's year-round demand for NFC must have substantial juice storage capacity. Initially Tropicana built gigantic walk-in freezers at its Bradenton plant to store freshly squeezed juice. The "tunnels," as they are called in the industry, are equipped with small, well-insulated doors that open onto long, narrow corridors, in which are stored two towering rows of slabs of frozen juice. Sheaths of plastic wrap separate each frozen piece so that individual ones can be removed with relative ease. These aboveground tunnels that protect Tropicana's frozen wonderland from Florida's oppressive heat are signposts of the company's high-tech, futuristic terrain.

While the tunnels mark the forward thinking that has defined Florida's rapidly transforming orange juice processing industry, they also take minds back in time. According to industry consultant Allen Morris, "The only way you can make a quality NFC is to freeze it." But Tropicana no longer stores most of its juice in frozen freshly squeezed slabs. The company began to investigate cheaper storage arrangements when its success as the top RTS juice provider in the late 1980s meant having to store more to ensure a constant supply of juice. Always at the forefront of pasteurized orange juice technology, Tropicana was the first to purchase the massive tanks that have given expression to a new hybrid industrial-agricultural landscape: the "tank farm."[11] The original tanks each held 250,000 gallons. Tropicana's first tank farm, which it built in the mid-

1980s, contained twenty-eight tanks with a total capacity of seven million gallons. In the early 1990s competitors began building similar farms, which were slightly smaller, holding a total of five to six million gallons. In 2004 Tropicana added several one-and-a-half-million-gallon tanks to its Bradenton farm, already home to twenty million gallons of juice divided among one-million-gallon tanks.[12]

The bulk liquid storage technology that all the big NFC processors now use was first developed at Purdue University for tomato juice. Apple juice was next before Tropicana became interested in adapting the technology to citrus. Morris notices a marked difference in Tropicana's juice: "If you had tasted Tropicana in 1991 versus now it wouldn't taste like the same quality . . . not even close." He attributes the decline to the fact that the majority of Tropicana's juice is now, unlike in 1991, stored in tank farms. He explains that, with aseptic tanks, the juice is heated before it goes into the tank and then again before it goes into the package, but with frozen block warehouses the juice is frozen and then heated only before it goes into the package. The more the juice is heated the more its freshly squeezed flavor is depleted. But one more heat treatment is not the only extra involved in storing juice in a tank farm. As the Tropicana director says, the juice is "pasteurized, deoiled, put into an aseptic tank farm, and then brought out, blended, add-back [added] and put into the package."[13] The only missing step is the final heat treatment that the juice usually receives before packaging.

The Web site of aseptic citrus system provider FMC FoodTech identifies the company's "core citrus capabilities," four of which are key to aseptic storage technology. The company provides: (1) aseptic sterilization systems; (2) NFCOJ deoiler systems; (3) NFCOJ deaeration systems; and (4) the technology for "aseptic juice blending and flavor addition."[14] Together these four systems perform the essential steps in storing NFC in tanks. The first sterilizes the juice with heat, the last blends back the flavor that is lost over the course of the aseptic storage procedure. Goodrich's explanation of the how and why behind the deoiling of pasteurized juice sheds light on the middle two systems: first, she says, "the oil probably contains the most unstable portion of the juice flavoring. When you have it in aseptic tanks for an extended period of time you want to have fairly low oil levels"; and second, "just the process of deaerating the juice, which you have to do to store it in aseptic tanks, you have to get rid of the oxygen, that again is a vacuum stripping process and that essentially de-oils the juice as well." Deaeration, she says, or stripping the oxygen, is necessary to prevent oxidation. An expert in agricultural technology who used to work at Tropicana says that deaeration "evaporates at high temperatures."[15] Aseptically stored juice is therefore by definition not "gently pasteurized," as the Simply Orange Web site claims.

Having heated, deoiled, and deaerated their juice for aseptic storage, pasteurized juice processors need the technology FMC FoodTech provides for "aseptic juice blending and flavor addition." As the Tropicana director says, there is no need for such faculties for juice stored frozen because it has "never been pasteurized, never deoiled, it's just raw juice that's frozen." There is no need to add flavor because "the flavor is still carried in the frozen block." Maybe that is why Morris thinks Tropicana's Pure Premium is not as good as it was when most of it was stored in tunnels rather than in tanks. Not everyone is so forthcoming, however, about the difference in quality between the fresh juice defrosted from the tunnels and the heated, deoiled, deaerated, and flavor-added-back juice poured from the tanks. When asked which is better, the agricultural technology expert who worked for Tropicana at the time is careful not to be critical: "I'm not going to answer that question."[16]

This expert is more forthcoming when explaining why flavor addition is part of the aseptic storage system: deaeration "strips a lot of volatiles." How many depends "on what temperatures you're running it at and how much vacuum you're pulling." While deaeration, he says, is "a pretty standard procedure for anybody storing aseptically," how it is done is not. He says that processors "vary their product specifications in terms of temperature and vacuum" depending on the market. If, for instance, "they are getting a pretty high dollar for oil phase essence, then they'll run some more vacuums or more temperature or something to strip more volatiles." The specifications also vary to ensure consistency. As Goodrich says, processors sometimes take out more oil than usual so "you can then add back your desired oil flavored package." The practice is widespread, she says, because "it's easier to control the overall flavor if you can have a little bit of leeway with your flavor package." A microbiologist at Florida's Natural agrees, saying that when deoiling NFC juice for storage his company "removes more [oil] than we need to," so as to "standardize the juice." He echoes Goodrich: it is "better not to put oil in and then have to remove it."[17]

Company priorities are one factor determining the quantity and quality of the flavoring components lost during the aseptic storage process. The miles the juice travels also play a prominent role in the stored juice's flavor profile. As Morris observes, the new shipping technology that has made Brazilian NFC imports possible is also making it "cheaper to put orange juice on a ship than in a tank farm," but there is still "the quality issue of heat treatment." The problem with juice that travels is "every time you move [the juice] you have to heat-treat it."[18] So "gently pasteurized" is even less applicable to NFC manufactured overseas than it is to juice stored aseptically down the street.

Despite being subjected to more heat, Brazilian NFC is making a mark on the North American market. Daniel King, director of technical services at the Florida Citrus Processors Association, admits that the quantity being imported is significant. The Tropicana director confesses the company "is buying from Brazilian NFC," but he says the amount is "tiny," making it nothing more than a "hobby." He does not speak directly about his company's adversary, though he does say that Cutrale, to whom Coca-Cola has contracted out most of its processing, makes NFC in Brazil that is sold out of a New York tank farm. Much of this juice likely is marketed under the Coca-Cola Simply Orange NFC brand label.[19]

From these developments we know that 1) any NFC product with widespread and year-round distribution inevitably contains some quantity of aseptically stored juice. Tropicana's successful 2005 bid to increase its aseptic storage facilities to hold more than 110 million gallons of juice suggests the proportion is growing; 2) national brands are adding Brazilian NFC to their blends; and 3) aseptically stored juice is aggressively heated and handled, the Brazilian imports more so.

The Simply Orange TV ads likewise distort reality. One depicts birds happily chirping around an orange tree in a Garden-of-Eden-like setting. The picture intimates a product that will magically take its consumer to tranquil, untouched nature. But the taste of paradise that a sip of Simply Orange promises is irreconcilable with the geography of the groves where juice oranges grow. Although the focal orange tree in the ad grows freely in spacious greenery, the number of commercial orange trees planted per acre is trending upward, from an average of just over 80 per acre in 1972 to more than 130 in 2002. Grown via grafting, and planted methodically in rows that in some groves cover as many as forty-five thousand acres, the trees lose their individuality. Each is a union between a chosen scion, or variety, and rootstock that is selected to suit environmental, production, and scion demands. About 80 percent of all commercial orange-bearing trees in Florida grow on five rootstocks. Frederick Gmitter, a citrus breeder and horticulture professor at the University of Florida-Gainesville, says that the growing of processed juice oranges in Florida is "basically a monoculture system."[20] In most groves each tree, except for the odd mutant, is identical to its neighbor.

Agricultural terminology has not caught up with modern farming practices. The late Thomas Mack, founder of the Thomas B. Mack Citrus Archives and former professor in the citrus department at Florida Southern College, liked to talk about the "citrifacts" that he documented and compiled in a book of the same title. One of his favorites was the use of the word "grove" to describe commercial orange plantings. Mack read the definition in *Merriam-Webster's Dictionary*, which says that a grove is a "woody group of trees planted or growing naturally," while an orchard is "a planned planting or collection of fruit trees arranged for cultivation, and often enclosed." The rows upon rows of manicured orange trees that run alongside Interstate 90 in Florida are the definition of orchard, not grove. In his book Mack theorizes that the use of the word "grove" in association with Florida orange trees originated in the 1790s, when settlers came across groups of citrus trees growing untended along lakes and rivers. Mack writes that the plantings, located where Indians made their campsites, were assumed to have grown from human-scattered seed.

The settlers began to enclose the groups of seemingly wild trees, but even as they began to manage them the word "groves" stuck.[21]

The Swingle Citrumelo, a cross between a Duncan grapefruit and trifoliate orange, can take much of the credit for the highly structured geography of Florida's orange "groves." Renowned citriculturist Walter Swingle developed the rootstock in 1907, and the USDA re-released it in 1974. The hybrid's tendency to grow small trees enables growers to plant them more tightly, a trait that is attractive to an industry geared toward maximizing production. It became a rootstock of choice at Tropicana, which monitors its supplier groves to ensure they will meet the company's juice needs.[22]

Growers Rory and brother Robbie Martin remember when trees were spaced thirty feet by thirty feet and how, as young boys, they weeded their grove themselves. Not anymore. Swingle's release partly explains the almost 50 percent reduction in tree spacing that is now typical. Weeding is now done by herbicides, Robbie says. He points out that the trees are now planted in a line that "takes out gaps between trees so [they] grow like a hedge in one solid row." The resulting melding of treetops into one massive canopy generates the humidity that increases the incidence of disease, and also the need for pesticides and the corresponding delivery machinery that is nowhere in the vicinity of the freestanding, made-for-TV, Simply Orange tree.[23]

Jim Brewer, a midsize grower under contract with Tropicana, has expanded his grove over the last thirty years from 150 to 3,000 acres. He says that the trees are now spaced twenty-five feet by twelve and a half feet, dimensions that translate to 140 trees per acre, producing twice as many trees per square acre as the "old style of planting" that was the norm when he was growing up.

> The idea for [the close spacing] was . . . like anything else in farming, because we're all raised and living today, the faster we produce something the more money comes back the quicker and the quicker . . . so we burn them out so to speak. A tree that used to last to be eighty, ninety, a hundred years old or even older couldn't make it today because we give them such good fertilizer. . . . And they're sprayed and they're watered right and they're cared for and so they produce. Like a cow, like growing cattle, there are only so many calves you're going to produce with a cow. If you have her having the calf every year, every year, every year, she won't last as long as that cow that has one every 1.5 years or something like that, [one] that takes her time and drags along. So she's just going to die quicker.

Although most growers say that the older and more ornery the tree the sweeter the fruit, the elderly respond too slowly and unpredictably for these high-producing communities.

Brewer envisions the future of the Florida orange grove as a twenty-year crop. "We've all decided that what we'll do is we'll plant more trees per acre and we'll burn them out. We'll get whatever's in them, get them out and replant them. A grove will be a twenty-year crop, or maybe twenty to twenty-five, and we will push that grove up and plant a new one."[24]

Jim Griffiths acknowledges that the twenty-year grove is a possibility because, he says, when "the trees grow together [in a hedgerow] there is a tendency for yield to be reduced so they maximize at probably twelve to fifteen years today. Likely after that, unless they're managed properly, their yield will come off that peak." Nevertheless, he is not convinced the twenty-year grove is the best solution to securing high rates of production. He compares the hedgerow to the wider set trees he saw while touring groves in the 1950s. The latter "continued to increase production until they were forty to fifty years old." He intimates that until agricultural researchers take the time to perform the long-term experiments that are needed to determine the appropriate way forward, quick fixes, like the one Brewer mentions, will hold sway.[25]

A University of Florida professor who works at the Citrus Research and Education Center says that the innovative and slow producing research that Griffiths calls for is unlikely, as citrus researchers are "just doing what they have to do to get by, to get the next pay raise. It's called punished by rewards. There's a book about that, that talks about what happens to people when they're in a system where what they're working for is the next pay raise or the next promotion . . . how that hurts really being innovative, and creative and thinking about new ideas, new things, things that are really big improvements." The professor speaks from experience: "The system is set from headquarters where it's teaching and research. Those guys are in the same situation, they've got to get promoted. It's like a military check-off list when I get evaluated. [There is] like a five page list of check-off stuff. That's what they do in the military and the military I don't consider to be very inventive at all."

The tenure system is not the only obstacle that the professor sees to innovation: "Funding comes into this as well because people think they want to know what the results are going to be before you do the research."[26] The experiment that takes years to generate results—and this encompasses almost every trial using perennials such as citrus—has trouble attracting support. Griffiths's description of the current situation is similar: "We don't have good experiments going on now, we have opinions." Referring to the example of high-density planting, he says, "We do things as matter of art not science in hedging and topping. Each guy has his own idea and somebody's right but everybody isn't."[27]

Citrus nursery owner Roland Dilley is more scathing in his criticism of Florida's citrus academy: "You take research and PhDs, a lot of them have a one track mind. They don't want to share anything, but 'I want to write the paper, that's why I'm here, I'm going to write the paper.' So they write the paper and get credit for it whether it works or not and there's just pile after pile of stuff that I've kept that didn't work." Dilley is especially unimpressed by the work of the CREC, a mecca of Florida citrus research. He says bitterly: "I'd just as soon they'd close the damn doors for what you get out of them, the money that's spent."[28]

The orange tree that Simply Orange represents as a respite from the pressures of human activity is, when placed against the ailing and time-pressed surroundings in which juice oranges truly grow, a honey-coated dose of irony. The "circuit," as Brewer calls it, of orange grove turnover schedules a nonstop workday for each tree in the orange grower's army. These trees, under a strict regimen that includes treatment with noxious sprays, do not, as does the Simply Orange tree, welcome spontaneous and leisurely visits by cheerful pollinators of fruit blossoms. The trees that shoulder the oranges for Simply Orange's product are equally if not more stressed than the viewer to whom Simply Orange ads offer a reprieve.

The marketing of Tropicana's Pure Premium NFC orange juice is no less deceptive. Just off of an interstate highway at the intersection leading to Tropicana's largest processing plant in Bradenton, a 2004 billboard read: "Nothing added, almost nothing removed." A picture of a carton of Tropicana Pure Premium introduced the signature statement, and an image of a coiled orange peel completed it. The drawn out motto is misleading; much has been removed and added, including the peel that the ad depicts as the only part of the fruit that has been removed.

The presentation of a highly processed product as pure and simple is an old advertising strategy. The familiar image of an orange punctured with a straw that is the hallmark of the Tropicana Pure Premium label is a throwback to the processed food ads of the 1920s and 1930s. In 1929 *American Magazine* ran a Campbell Soup Company ad that illustrated a tomato soup can bursting through a ripe tomato. Advertising historian Roland Marchand describes the ad: "Against the back-drop of radiating beams and nostalgic country scenes, Campbell's offered urban readers an image of miraculous transformation rather than an education in actual production processes." As Marchand observes, the ad and ones like it offered "therapeutic resolutions . . . for such core dilemmas of the society as the challenges of . . . 'overcivilization' to the simple virtues of nature."[29] The gap between the conception and true composition of processed orange juice may be a sign of a society once again overwhelmed by life-changing technology.

During the 1961 hearings, major orange juice processors presented themselves as passive producers with no interest in controlling consumer consciousness. Since the hearings companies such as Coke and Pepsi have showed the ease with which they can shape consumer perceptions through language and

image manipulation. Tropicana's unleashing of "not from concentrate" was a feat of regulatory circumvention and product
reinvention. Throughout the hearings processors criticized the
FDA's proposals to change the names they gave to pasteurized
orange juice. After arguing that reform constituted counterproductive, confusion-generating action, these same processors orchestrated a new name, "not from concentrate," for an
old product. As NFC has evolved ever further from fresh, the
name, intended to denote something close to just-squeezed,
has remained the same. Little has changed since the 1961 hearings. Most orange juice consumers are as unaware of exactly
what they're drinking now as they were five decades ago.

NFC's popularity is a testament to the standard of identity's weakness as a tool for regulating marketing behavior. The
next chapter, which examines the modern flavor pack and its
widespread use in orange juice, underscores another failure of
standardization: its inability to keep pace with advances in food
production technology.

XIII
Fabricating Fresh

Traditionally the flavor of processed orange juice depended only on the oranges squeezed. Now the flavor is sourced from all parts of oranges everywhere. Many consumers would be shocked and disappointed to learn that most processed orange juice, a product still widely perceived to be the definition of purity, would be undrinkable without an ingredient referred to within the industry as "the flavor pack." The fact that the modern orange juice flavor pack has retained almost complete anonymity is a symptom of orange juice standards of identity that were literally "fixed" pursuant to s.341 of the FFDCA in the early 1960s. Significant advances in orange flavor manufacture have rendered the original regulations, upon which the current rules are based, out of date. The growing complexity of the orange juice flavor pack, and the ambiguity surrounding its regulation, highlight a fundamental problem with the standard of identity. Its content rather than its packaging-focused approach is ill-suited to the regulation of a processing sector that has become sophisticated enough to skirt the FDA's radar.

During the 1961 hearings orange juice processors sur-prised the FDA with testimony that they had been experiment-ing with orange essence since the 1950s. After the hearings James Redd became known as the man who turned orange flavor ex-traction into a highly lucrative commercial enterprise. Redd, who also flew to Brazil after the Florida freeze of 1962 to lend his expertise to Brazil's growing juice-processing industry, started Intercit, the first company in Florida devoted to recov-ering orange essence for resale to juice processors. Firmenich eventually bought Intercit, which is now home to the interna-tional flavor and fragrance company's Citrus Center in Safety Harbor.[1] Renamed, and the essence recovery systems that Redd engineered for Intercit updated, the Citrus Center retains the same purpose: to capture the essence of orange juice.

The deal that now almost every orange juice processor makes to create a decent-tasting processed juice involves an in-tricate give and take between orange juice processor and flavor manufacturer, usually an outside flavor and fragrance house operator. Typically, the orange oils and essences that juice con-centrators collect during evaporation are sold to flavor manu-facturers, who then reconfigure these by-products and sell them back to juice companies. The flavor house that purchases this material may use some of it for fabricating fragrances and flavors for other products, and the rest is broken down and reconfigured into "flavor packs" for reintroduction into orange juice. Renée Goodrich of the Citrus Research and Education Center says that "it's only been fairly recently that we've had the good equipment to take out excess peel oil that lets us get down enough to then add back the flavor packages."[2]

While most brands rely on flavor packs to restore to their product the aroma of freshly squeezed juice that is destroyed during processing, others use the flavor pack to imitate the best imitations. Former flavor house employee Daniel King, who moved on to become director of technical services at the Florida Citrus Processors' Association, recalls customers asking for blends that, when reintroduced into their juice, would mimic the taste of a specific brand. Minute Maid has a reputation for having an especially distinct flavor. Its "from concentrate" orange juice, says the former agricultural technology expert at Tropicana, is known for the "flowery flavor package that's floating on top." His colleague at the time, the Tropicana director, is more specific: "If you drink Minute Maid it has and always has had a unique candy type flavor to it. The oranges as they come out of the grove aren't always like that. But obviously the orange has that flavor and that's one that is highlighted in whatever they're adding to their juice. Maybe people who like Minute Maid like it for that reason."[3]

Not all orange juice companies can afford distinctive flavor packs. King says the blends of different companies vary in their degree of sophistication: the "more specialized the flavor package is . . . the more expensive components one blends into it . . . the more complexity the cost."[4] But however modest the investment, it appears to be worth it. Industry consultant Allen Morris has tried undoctored concentrate: "If you taste the bulk concentrate that hasn't had the essence added back it just tastes like sugar." Tropicana's former agricultural technology expert offers a parallel perspective, particularly for frozen concentrate, and the reconstituted cartons of orange juice that Minute Maid makes from it: "Once you strip all your volatiles out . . . what have you got? Brix [orange sugar solids] and acid. [You] cut it back and add a flavor pack until you've got orange juice."[5]

During the 1961 hearings juice processors vigorously denied suggestions that their interest in essence derived from its capacity for deception. They did not want to give the FDA or consumer representatives the impression that they were using essence to cover poor-quality juice. However, the contemporary flavorist is not shy to admit that the modern flavor pack serves, in the words of the Firmenich flavorist, "a protective or masking role." The high-tech concoctions of varying fractions of orange essence and oil soften the effects that processed orange juice suffers at the hands of processors who, the flavorist says, like to "crank up" the heat to get rid of dangerous bacteria and increase shelf stability. Goodrich does not think the smaller-scale premium juices made by companies such as Odwalla contain flavor packs—more evidence that they act as a Band-Aid for processors who are rough with their oranges.

The Firmenich flavorist says that the flavor pack represents "your signature, that's what makes or breaks you. If you get a good add-back [flavor] package you can do justice to a lot of negative flavor qualities." The flavor pack ranks high in his estimation of the key factors contributing to processed orange juice's salability: "It's really important, it's really what gives you the punch in your finished product."[6] In 1961 the FDA was concerned about orange essence's strength to overcome off-flavors. If it had anticipated how sophisticated flavor packs would become it might have devoted more than the last few days of the hearings to the regulation of essence and oil reintroduction.

Russell Rouseff, who has been researching citrus flavors at the CREC since 1974, describes the unique challenge of reproducing the taste of freshly squeezed orange juice: "The balance of orange particularly is a very interesting thing. Whereas in other food products we can find one or two compounds in [each] that, when you smell it, the first thing you think of is that product . . . The interesting thing in orange is . . . [that none of the compounds that contribute to the orange's aroma] individually smell of orange. And so orange flavor is just a combination of a lot of different flavors." He can list the one or two chemicals that define the flavor of bell peppers, grapefruits, apples, and peaches, but the aroma of an orange does not derive from any one single compound: "It's a combination of things and a certain balance that produces that aroma."

At Firmenich, those who break down orange essence and oil into its components say the same: "The oil consists of hundreds of chemicals. Citrus is a very, very complex thing because it's not just one or two chemicals that make it taste the way it does. It's this harmonious synergy that is occurring between some of the components that are in large concentrations and some that are in such low levels you wonder how they can do anything." Such complexity explains why, after years of research by Florida state government and the industry, neither has been able to bottle the aroma of a freshly squeezed orange. According to Rouseff, "Right now the formula for fresh [orange] flavor is just about as elusive as the formula for Coke." Only not even Coke has the formula for this one. Flavorists say that the most easily isolated components don't result in the best-tasting juice, and using them all would be prohibitively expensive.

The flavor of fresh orange juice is, in Rouseff's words, "the holy grail of most flavor people." The flavorist at Firmenich agrees that the taste of fresh orange juice is the gold standard that every processor tries to emulate. "If I came up with that solution I could be sitting on a beach somewhere talking about it. It would be monumental," he says. The ex-Tropicana agricultural technology expert, who is also familiar with the inner workings of a flavor house, says that improvements in analytic methods are helping flavorists to "really tear the oils in general apart into individual compounds," adding that "the guys at Firmenich could show you some mind boggling chemistry."

The subjective nature of taste adds more layers to the highly sophisticated science of flavor that was developing around the time of the 1961 hearings. The flavorist's understanding of both orange essence and consumer preferences has advanced to the point that chemists at Firmenich have been able to determine that ethyl butyrate is the orange constituent that North Americans cherish most, albeit unknowingly. The chemical imparts what North Americans like best about the flavor of freshly squeezed orange juice. But the preference is not universal. The Firmenich flavorist says that in order to satisfy the palates of other nations, such as Brazil and Mexico, other chemicals, for instance the "decanols," might be emphasized.[7]

The wall of secrecy that surrounds orange juice flavoring materials, and that concerned the FDA in the early 1960s, still marks the flavor side of the orange juice industry. Jim Griffiths says he does not have "any good data" about whether flavorings do a better job than cut-back in restoring fresh flavor to processed juice because: "We don't generally know, everybody has hidden what they know about flavor. And whether it be Firmenich, or Mastertaste, or Givaudan, or whomever, they all got their own ideas." He says that Coca-Cola is experimenting the most with flavor but that he is unsure about the extent of Tropicana's efforts in this area. "Over the last seventy-five years, Coke has been working at flavor in whatever they're working on. And they have got a hell of a lot of high-priced people." When asked whether Tropicana produces its own flavors, the Tropicana director's answer is a simple "no."[8]

Rouseff, a state employee who has tried to penetrate the

walls of the private sector, notes that "I'm from the academic side and sometimes company people are very, very jealous about their flavor approaches simply because that is such a market-differentiating feature to their product. They guard that very judiciously."[9]

The mystery of fresh orange flavor that Firmenich and its competitors are trying to decode extends to FDA regulation of what the flavor companies have come up with thus far. FDA orange juice flavor regulation begins with the orange juice standard of identities. Although the 1961 hearings made clear the FDA's reluctance to approve a substance about which it knew little, the agency ultimately approved orange essence as an FCOJ additive. The standard of identity allows "orange oil, orange pulp, orange essence (obtained from orange juice)" and "water," among other ingredients, to "be added to adjust the final composition."[10] Presumably the rationale for allowing the addition of essence and even water to concentrate is that both are lost during concentration.

The FDA's stance on the addition of orange essence to not-from-concentrate, or pasteurized orange juice, is a separate and somewhat unresolved issue. The standard of identity for pasteurized orange juice says that "orange oil may be adjusted in accordance with good manufacturing practice." No provision is made for orange essence addition. The lack of clarity as to essence's acceptability in NFC has generated disparate views within the industry. King says, "If you read the federal rules, the standards of identity, one [reading] basically says if it comes from an orange it can be put back into an orange. Others believe that's a liberal reading of the rule and don't do so because they don't believe the rule allows it. The regulatory agencies are not in a position to address that. There are those who readily add back [water-soluble orange essence] as a flavor component and those who don't. I don't think there are any clear-cut answers. . . . There are those that feel when adding natural aroma you're adding water and the statutes clearly do not allow you to add water. Others say natural aroma is a natural by-product of the process and therefore should be allowed."[11]

The senior research and development manager at Florida's Natural says that his company does not use water-soluble essence in its NFC, adding that "there was a time when it was considered dilution but it isn't considered dilution now." King doesn't seem to agree. He says that the FDA might still disallow the addition of "aroma," which is his term for water-soluble essence, to NFC, because "you are adding a water-based material . . . and you might imagine that if you are adding back a very weak aroma at a very large volume . . . you are now creating juice that never occurred. That's part of the argument against it, is that you can go against the spirit of the law." Tropicana, like Florida's Natural, is equivocal with respect to whether water-soluble essence is an acceptable addition to NFC. The Tropicana director says, "[We] typically don't [add water soluble essence] . . . it has water in it, so you know you're diluting . . . but I don't know, maybe you can."[12] Neither he nor the Florida's Natural research manager will commit to either a conservative or a liberal reading of the law.

The ex-Tropicana employee, who is more familiar than the Tropicana director regarding the company's flavor technology, is no more definitive about the FDA's regulation of water-soluble essence. Still with Tropicana when interviewed, he evades questions as to whether Tropicana uses essence in its NFC. He does say there are no limits on water-soluble essence in the federal standards. He also asserts: "Anything that comes out of the orange during processing can be put back." This statement explains the testimony of Howard Trumm, of juice company Libby, McNeil, and Libby, during the 1961 hearings that deaeration, the process that strips oxygen from pasteurized juice to prevent oxidation during storage, also sucks out the essence of the orange. Now that most pasteurized orange juice is stored in tank farms, deaeration is common. Those who farm pasteurized orange juice thus have a defense, if not the authority of law, for adding essence to juice milked from the tanks: the practice is okay if whatever leaves the juice can be put back in.[13]

The Firmenich employee who synthesizes flavor packs for some of the largest juice companies says that he has not encountered any companies that request flavor packs without water-soluble essence. But he is aware of the concern that water-soluble essence dilutes NFC and guardedly says that he doesn't know whether adding essence to NFC is a common industry practice.[14]

Whether or not essence is added to NFC, no one doubts what the advantage would be. "The reason [water-soluble material is] important is because most of the volatile components that come off quickest are also the most soluble in water," says King. "So [water-soluble essence] may well be adding back the freshest characteristics."[15] Considering the boost that water-soluble essence gives to juice, it seems unlikely that NFC producers who say water-soluble essence is acceptable would choose not to use the taste-enhancing material.

Although King says that "the regulatory agencies are not in a position to address"[16] whether water soluble essence is a permissible addition to NFC, after a year of repeated queries the FDA responded that the agency approves of flavor packages made exclusively of orange oil. Because the FDA does not state otherwise, they can presumably be added back at any point in the process. Flavor packs containing orange essence face more restrictions, partly because of the FDA's definition of orange essence. The agency does not define orange essence in the orange juice standards of identity, but Catalina Ferre-Hockensmith, a staff member of the FDA's Center for Food Safety and Applied Nutrition, says that it is "the volatile flavor/aroma fraction recovered when concentrating orange juice that contains the essential flavoring and aroma constituents from the juice." She adds that "it contains fairly high levels of alcohol and little, if any water." Because by this definition essence is the mixture of volatiles that escape during *concentration*, the FDA restricts orange essence to products that have been concentrated. It is an optional ingredient in the standards of identity for frozen concentrated orange juice; concentrated orange juice for manufacturing; canned concentrated orange juice; reduced acid frozen concentrated orange juice; and concentrated orange juice with preservative. The standards of identity for pasteurized orange juice and orange juice from concentrate do not, Ferre-Hockensmith says, "provide for the direct addition of orange essence," though essence can be added indirectly: "These standards provide for the addition of frozen concentrated orange juice, which provides for the addition of orange essence."[17]

Regarding the company's use of oil phase essence, the Tropicana director reports that "oil phase seems to be the predominant essential oils and essence that you add back to NFC." Moreover, he adds, "You can add oil phase back to a from-concentrate [otherwise known as reconstituted concentrate, or recon] product, and a lot of people do." He says that timing does not matter: "You can dose from-concentrate before or after pasteurization, you can put the flavor back in after pasteurization or before pasteurization."[18] But according to FDA's Ferre-Hockensmith, neither not-from-concentrate nor from-concentrate can be "dosed" with flavor other than that derived from the peel. If flavor packs containing essence, whether oil or water phase, are added back to from-concentrate, the step cannot be performed after the reconstituted orange juice has been pasteurized. Such flavor can be added back only to the concentrate from which the final product is reconstituted.

While acknowledging that his company adds essence back to NFC, the Tropicana director will not divulge when it does so.[19] But, according to the FDA, the addition of any essence independent of concentrate, whether water or oil phase, breaks the rules. The standard of identity for pasteurized orange juice does not make allowance for the fact that deaeration, which has become prevalent in NFC manufacture, strips juice of essence. Until it does, NFC producers must keep their flavor add-back systems simple: peel oil only is acceptable.

Renée Goodrich, who as an employee of various juice processors worked with the FDA's orange juice standards of identity, acknowledges that "there are so many gray areas in food law." For example, the wording of the law is unclear regarding the question of whether the essence and oil returned to processed orange juice has to be derived from the same oranges squeezed. Goodrich notes: "If you read the letter, the actual rule, you would get the impression that you have to make [the flavor] from the oil that you took out from that orange." King agrees: "It seems to me to add back it had to come basically . . . from the fruit involved. If we processed Brazilian fruit here then those oils and aromas as such could be added without further labeling. I believe—now this is a reach—that if you bring in an outside source of oil or flavor component it must be labeled as such." Although King readily acknowledges that his "familiarity and memory of labeling law is less than perfect," he does seem to feel strongly that processors are mistaken if they think the flavor they add back to their juice is okay as long as it is extracted from *Citrus sinensis.* "The implication is that it has to come from the fruit being processed as opposed to somebody else's fruit being processed," he says, recalling that "there was some issue doing that in the United States." But as to whether a juice marketed "made in Florida" can contain oil derived from the Pera orange, a sweet orange variety grown in Brazil but not Florida, King says, "I just can't remember what the law stated."

Mike Sparks, former consumer and market researcher for the Florida Department of Citrus, also doubts the legality of adding non-Florida products to juice sold as 100 percent Florida-squeezed. "I'm pretty confident that our internal controls and the inspectors would not allow any foreign oil," he says.[20]

The FDA's unwritten policy toward the origins of the oranges from which flavors are extracted resolves two misunderstandings that Goodrich, King, and Sparks all seem to share. "The natural orange flavor added to 100 percent Florida juice does not have to be derived from Florida oranges," says Ferre-Hockensmith. "We have no provisions in our juice standards of identity that stipulate from what oranges natural orange flavor must be derived. Therefore, the natural orange flavor added to 100 percent orange juice does not have to derive from the orange squeezed."[21]

Goodrich's description of how flavor packs are put together may explain why the FDA is not stricter on this issue: "Oranges are grown worldwide. And we get some of the flavor components from Brazil. You'll see barrels of cold-pressed oil from Brazil in all of the flavor companies' parking lots. So how you keep the traceability of that is a big challenge. You can probably verify that it's *Citrus sinensis,* but, after that, once it's all blended together, it's hard to figure out how you could keep track of whether there was a component from Indonesian cold-pressed oil."[22] The FDA's tolerance of flavors in orange juice from oranges other than those squeezed is pragmatic. It does not demand a greater level of purity than it can enforce.

The FDA's lack of restrictions on the geographic origins of the natural flavors used in orange juice is not a benign concession to processors and their customs. The flavorist at Firmenich calls attention to the minimal control his company has over the raw materials it receives: "There's things we're trying to do to impact the industry from the flavor standpoint that, we only have so much leverage because we can't go to the growers and say don't apply the pesticides because they're showing up in our oils. We can't do that because it's a by-product. They're in the business for juice, that's where they make their money, not with this oil, we make our living off the oil." He admits that Firmenich is a global company that is based in Geneva, "so we buy from the Brazilians, we buy from Costa Rica, we buy from Israel, we buy from California, we buy from Florida." But the quality varies. He compares Valencia oil, which is derived mostly from Florida oranges, to Pera oil from Brazil. "The different chemicals that we have identified as [impacting flavor] come in different levels. The Pera doesn't seem to have what we call the goodies in the same concentration as the Valencia. So Pera oil on the market is lower." Valencia oil, he says, is the benchmark. Apparently Florida, which has a reputation for growing a Valencia like no other, also has a reputation for producing the highest quality oils.

But quality does not drive all of Firmenich's decisions; cost is also an important consideration: "If we have the ability to maximize our purchasing power and buy Pera oil that's less expensive than Valencia yet make our products taste as if we were using Valencia oil, then that's where [we're going to get the oil]," he says. The flavor engineer will not divulge whether Firmenich buys more oil from Brazil than from elsewhere, but he hints that Brazil is a large supplier: "We make decisions based on our global entity, so we . . . buy more Brazilian oil than, say, some of our competitors that are not as global as us. . . . It's due to the strategy of the company, what really they're looking for." This likely means buying raw materials from the lowest bidder, so presumably much of the juice packaged in Florida contains flavoring material originating elsewhere. The flavorist lacks conclusive data, but, based on "some snapshot analyses of some products over time," he concludes that the regulations, such as those relating to the quantities and types of pesticides used to grow oranges, "are not adhered to as much globally, maybe as much as they are here" in the United States. According to analyses of the oils, he says, "We seem to see more things [in oils coming] from places other than Florida."[23]

An unclear relationship exists between the country of origin and pesticide inputs and flavor by-product output. The flavorist cautions that "it's hard to make that assessment because we're only capturing a few points, and we might have only caught a real bad field at one time and that only represents 0.00005 percent versus the other 90 percent of the crop is fine, so it's hard to really say."[24] But by letting orange extracts from countries with unfamiliar, perhaps even nonexistent, pesticide regulatory regimes into U.S.-marketed orange juice, the FDA is subjecting consumers to the risk of consuming potentially dangerous and unmonitored chemical residues.

Although orange juice processors won FDA approval to add to their juice orange essence and oil from oranges other than those squeezed, the FDA's position is less clear on whether

the flavoring materials added to orange juice have to be labeled. Ferre-Hockensmith relates the agency's general rule for
labeling "ingredients": "When a packaged food is made using
two or more ingredients, each ingredient must be listed by its
common or usual name on the label in descending order of
predominance by weight." Flavors are an exception, as their
multichemical makeup can be collectively listed simply as
"artificial flavor" or "natural flavor." Though the chemicals
making up a flavor need not be labeled, if flavor is added to a
product it must be listed as an ingredient: "Standardized foods
such as orange juice are not exempted from ingredient declaration. Therefore, all ingredients used to make a food such
as orange juice (including flavors) must be declared on the
label."[25] Since passage of the 1990 Nutrition Labeling and Education Act, standardized foods, which prior to the Act could
sport labels that were silent with respect to contents, no longer
receive special treatment.

Ferre-Hockensmith's inclusion of flavors as ingredients
is similar to FDA official Lowrie Beacham's position during the
1961 hearings, where he identified essence and oil as "ingredients," not "natural constituents," of orange juice. However,
"flavor" is typically not listed as an ingredient on the label of
100 percent orange juice products. The discrepancy can be explained by a ruling the FDA published in 1993 in response to
an industry attempt to convince the FDA that orange oil and
essence fit under the definition of a "processing aid" in Federal
Code Section 101.100(a)(3)(II)(b): "substances that are added
to a food during processing, are converted into constituents
normally present in the food, and do not significantly increase
the amount of the constituents naturally found." The section
exempts substances deemed processing aids from the general

requirement that all mandatory and optional ingredients of standardized foods appear on the food label.

The FDA was not persuaded by the industry's interpretation. Its 1993 ruling clarifies that the Section 101.100(a)(3)(II)(b) exemption pertains to incidental additives, which are defined as: "substances that are present in a food at insignificant levels and that do not have any technical or functional effect in that food." The ruling declares that orange oil and essence do not qualify as "incidental additives" under the exemption because oils and essence are added to fruit juice to accomplish specific technical functions (such as achieving "uniform quality and organoleptic properties"), and "the level of use of the added ingredients is not insignificant."[26]

During the 1961 hearings, FDA representatives debated whether to allow the addition of orange essence and oil to orange juice because of their concern about the naturalness of the flavors then in use. Processors reassured the FDA that the essence and oil they reintroduced were identical to what was found in raw juice, and so the FDA concluded that they could be added and need not be labeled. However, the flavor packs that are mixed into today's commercial orange juice products are different from anything nature created. Their uniqueness is reflected in the FDA's new mindset. It no longer agrees with Lowrie Beacham's 1961 conclusion that the addition of orange flavoring extracts "is not in and of itself, a very significant fact."[27] The FDA recognizes that orange oil and essence are critical to the transformation of an insipid liquid into something fragrant and fresh tasting.

The FDA's resolution that orange oil and essence are not processing aids for the purpose of the Section 101.100(a)(3)(II)(b) label exemption does not mean the additives must be la-

beled de facto. In the same rule that denied orange essence and
oil designation as processing aids, the FDA resolved: "When
these naturally occurring constituents (oil, pulp, and essence)
are removed from the juice for purposes of efficient processing, and these constituents, from the same types of fruit provided for by the standard . . . are added to the juice later in the
manufacturing process, there is no need to declare these constituents as 'ingredients' under section 403(i)(2) of the [Nutri-
tion Labeling and Education] act."[28] The rule runs counter to
Ferre-Hockensmith's reference to natural orange flavor as an
"ingredient" that must be labeled. Apparently the FDA accepts
that oil and essence are sometimes naturally occurring constituents rather than ingredients requiring labeling. The unresolved question is, when does a natural constituent become
an ingredient?

At the hearings Beacham testified that oil and essence
added to orange juice should be designated "optional ingredients" when they come from oranges other than those squeezed.[29]
The FDA's current position is not so clear. While providing for
a "naturally occurring constituents" exception to labeling, the
FDA explains: "Only if pulp, oil, or essence are added at levels
significantly in excess of those found in the raw juice from
which the juice is derived, or if they are obtained from an extraneous source, i.e., from sources other than the fruit from
which the juice is obtained (e.g. produced synthetically or purchased through a flavor supplier for artificially adding flavor or
texture to the juice), would they be ingredients subject to the
ingredient labeling requirements for standardized foods, as
set forth in part 130."[30] There are thus two instances when the
FDA would deem orange essence and oil to be ingredients
and therefore subject to labeling: (1) if they are found "at lev-

els significantly in excess of those found in the raw juice"; or
(2) "if they are obtained from an extraneous source."

This attempt at clarification fails on a number of counts.
First, the FDA's resolution that essence and oil are "ingredients" at levels "significantly in excess" of those found in raw
juice is hard to interpret given the context in which flavor is
added back to orange juice today. Essence and oil are most frequently restored to juice through flavor packs that look nothing like the natural essence and oil from which the packs are
assembled. "These flavor houses actually go in and fractionate
these oils into various sections and individual components
and then reblend from them," says King. "So you may put back
into the product a blend that is relatively unknown in nature."[31]
Few in the industry would deny that the modern flavor pack
bears no resemblance to the "natural" constituents of fresh
juice.

Moreover, the absolute quantity of individual volatiles
that are returned to the juice vary from what was initially
there. The flavorist at Firmenich acknowledges that "in the
case of NFC someone could have the argument that no, we're
adding back more stuff than what was there because we
shouldn't have taken anything out of it, we weren't concentrating the juice. . . . You're also doing that on the FCOJ. You
know we may be adding flavor systems that have much more
[ethyl butyrate] than what is naturally, normally there."[32]

Even the Tropicana director freely admits that the flavor
material going into Tropicana's juice is different than what
leaves it. He first became aware of Tropicana's use of "oil impact systems" in the early 1990s when the company began
keeping the majority of its juice in aseptic storage tanks. "You
had to deaerate the [juice] and then in order to package it and

get good flavor, particularly in the off season, you had to put [the flavor] back," he says. "And then what you put back could either be exactly what you took out or you could improve it. Of course the flavor companies, they're in the business to improve it. What they do, just like an oil refiner, they fractionate it out and strip off the goods and the bads and separate them and sell it back to you."[33] There is a consensus that processors hire flavor houses to improve upon, not merely imitate, the juice as it was in its raw state.

Given the rise of the twenty-first-century flavor house, the FDA needs to be more specific about whether the typical orange juice flavor pack is an ingredient or a natural constituent of the juice. Its statement that oil and essence are ingredients for the purpose of labeling if they are added "at levels significantly in excess of those found in the raw juice from which the juice is derived" is of little guidance. The determination does not, for instance, resolve whether a flavor pack containing individual chemicals, such as ethyl butyrate, in excess of those found in nature, should be treated as an ingredient to be labeled.

The FDA's declaration that essence and oil must be considered ingredients if derived from an "extraneous source" is equally equivocal. Its definition of an "extraneous source" as "sources other than the fruit from which the juice is obtained" seems to say, as Beacham did during the 1961 hearings, that essence and oil from outside oranges are ingredients, not natural constituents, of orange juice. As Goodrich noted, it is "highly unlikely if you actually understand how flavors are put together" that the essence and oil returned to orange juice ever derive from the same oranges used to make the juice.[34] Since most flavor packs are technically from an "extraneous source," the letter of the law says labeling is mandatory.

But rules are not always read literally. The FDA provides parenthetical examples of what it means by an extraneous source, or sources "other than the fruit from which the juice is obtained." To give processors the benefit of the doubt and assume that the examples are exhaustive, only those flavors that are "produced synthetically" or "purchased through a flavor supplier for artificially adding flavor or texture to the juice" would have to be labeled. This conservative interpretation of an "extraneous source" still raises questions. The flavorist's rearrangement of the chemicals in orange essence and oil into new and improved configurations would lead more than a few to conclude that the resulting packs are "produced synthetically." Alternatively, the very fact of buying a flavor pack from a flavor house for use in juice could be said to be the equivalent of purchasing "through a flavor supplier for artificially adding flavor . . . to the juice." In other words, the externally sourced flavor packs that are commonly blended into juice by definition "artificially" add flavor and therefore must be labeled. Given that orange flavor is rarely, if ever, listed as an ingredient in orange juice, processors must be interpreting the words "synthetically" and "artificially" more narrowly. The correct interpretation remains a mystery so long as the FDA does not define "synthetically" and "artificially" precisely.

The FDA's reason for differentiating natural constituents that need not be labeled from optional ingredients that must be labeled provides a clue as to when the agency expects orange juice manufacturers to label added flavor. It states: "When the constituents of the juice are added to the juice, in the same proportions as in the original juice, the basic composition of the juice is not changed." It continues: "It may be misleading to consumers if these constituents are listed in the ingredient statement of the juice when their addition has not altered the basic composition of the juice. Consumers might be led to believe that the juice has been fabricated from the individual juice components, or that juice contains higher levels of these constituents than would be the case if the ingredients had not been added."[35] In other words, when constituents of the juice are used to restore rather than modify a juice's flavor, the FDA does not believe they should be labeled. This is a reasonable conclusion but not applicable to the way orange essence and oil are used today. The FDA's evident concern that labeling flavors might mislead consumers is moot in light of the widespread agreement within the industry that the flavor packs mixed into today's orange juice contain, in the words of the Firmenich flavorist, "more stuff than what was there."[36]

Considering the industrialized nature of today's flavor packs, juice processors can no longer rely on the "natural" justification that they used during the 1961 hearings to win the FDA's approval. In representing the perspective of the processor of the early 1960s, Howard Trumm urged that "since [essence] is a natural mix of the constituents of the juice, restoration should be permitted without a label declaration." But almost never are the natural constituents of juice removed and restored true to form anymore. The big brands, with the help of flavor houses, essentially do "fabricate" their juice "from the individual juice components."

Chapter 14 tracks a change in the FDA's philosophy and methodology. The FDA is acting on its duty to protect the interest of consumers by paying more attention to the information that is sold with almost every processed food purchase. The shift, if sustained, could have major repercussions for food regulation, farmers, and society in general.

XIV

Moving Beyond the Standard of Identity

The processed food industry invested heavily in food technology during World War II, and in marketing after the war. It was beginning to reap the returns around the time of the 1961 orange juice standard of identity hearings, reinventing itself, and food as Americans knew it, along the way. All this forward action was taking place behind the scenes of a society that held fast to traditional norms that said women stayed at home. The testimony presented at the hearings was in keeping with the Anita Bryant orange juice ads that ran concurrently: "Mrs. Housewife" was in charge of the buying, preparing, and providing of orange juice for her family. Such social conventions, combined with the difficulty anticipating the power that the marketer and processor would acquire, explain the FDA's initial approach to food regulation. Its focus on food processes and contents, rather than on consumer education in executing its s.341 mandate to "promote honesty and fair dealing in the interest of consumers," made sense in the early days. The FDA was operating in a world in which the consumer was assumed to be a woman with a limited capacity to understand technology, and marketing was not yet almighty.

At the 1961 industry-dominated hearings, Faith Fenton was among a minority who favored FDA orange juice regulation. Yet she opposed the agency's choice of standardization. She was critical of the fact that when the FDA standardized a product, it moved the product's ingredient list from the label, where consumers could see it, into the Federal Code, where few would read it. Fenton had good reason to be concerned about standardization. In wiping ingredients' lists from food labels, it provided marketers with literally a blank slate to fill with reassuring words that continue to lull consumers into believing that the orange juice they buy is pure and simple.

David Kessler's tenure as FDA commissioner, from 1990 to 1997, led to a departure from the FDA's traditional content-focused approach to food regulation. At the annual convention of food and drug lawyers in Florida in 1991, Kessler told those in attendance of his aim to overhaul the food regulatory environment. In his address to the Food and Drug Law Institute at its annual conference in Washington in 1992, he reinforced his commitment to transformation: "I am here to talk about change."[1] While the FDA approached the 1961 orange juice hearings with the negative goal of reducing consumer confusion, Kessler brought to the FDA the more positive vision of informing the consumer.

In his remarks to the National Food Editors and Writers' Association in Washington in October 1991, Kessler lamented that "the FDA had reached the point where it was no longer viewed as a force to be reckoned with. The proverbial line in the sand—the line beyond which those regulated must not go—was drawn inconsistently, or it was not drawn at all."[2] The constant revival, during the decades that followed orange juice's standardization, of the debates that took place during the 1961 orange juice standard of identity hearings proves his point.

When Kessler assumed office one of his first and, within the food industry, more infamous moves was to act on the failure of the orange juice standard of identities to clearly establish and enforce the border beyond which processors and their promotions must not go. He chose Proctor and Gamble (P&G) as his target. After P&G refused to stop using the word "fresh" on its from-concentrate "Citrus Hill Fresh Choice" label, Kessler orchestrated a raid of the P&G orange juice operation. In the winter of 1991, just after having been sworn in as commissioner, he ordered U.S. marshals to confiscate twelve thousand gallons of the juice. Under authority of the 1990 Nutrition Labeling and Education Act, which deems "misbranded" or "mislabeled" products adulterated, he impounded the juice in a supermarket warehouse in Minnesota.[3] In the fall of 1991, before the National Food Editors and Writers' Association, Kessler explained his radical action: "Despite months of negotiations and years of warnings, some had not yet come to understand that the FDA regards the use of the term 'fresh' on highly processed products as false and misleading—and confusing to consumers."[4]

In 1961 processors admitted that the word "fresh" on a processed juice product was inappropriate. As mentioned in chapter seven, during the orange juice standard of identity hearings Tropicana vice president David Hamrick even apologized for a Tropicana label that described an impossible product: a "first fresh orange juice with unlimited shelf life." Hamrick preferred not to call the label "deceptive advertising," because, he said, there was no intent to deceive. But he did admit that the label could be construed as "misleading."[5] Back then there was no need for FDA intervention to remedy the situation. According to Hamrick, as soon as he became aware of the label he restricted it from further distribution.

The different food environment of the 1960s may explain why Hamrick, unlike P&G, acted voluntarily. In the 1960s the existence of pasteurized orange juice alongside dairy in dairy-shaped containers was enough to upset consumer representatives at the hearings such as Anne Draper.[6] She found unacceptable the complicity between manufacturer and retailer to put pasteurized orange juice in a place traditionally reserved for products that in the consumer mind epitomized "fresh." Tropicana's use of "fresh" for a processed product must therefore have stood out as an especially egregious breach of the consumers' faith in the integrity of their orange juice purchase, a ploy the juice processor could simply not get away with.

The tricks of the 1960s no longer worked in the more competitive and sophisticated food environment of the 1990s. As time has distanced the consumer from memories of home dairy deliveries, the subtle manipulation that Draper remarked upon is no longer visible to the average consumer or FDA official. Dairy is less a symbol of fresh, and its carton less a sign of purity inside. To convince consumers that store-bought refrigerated orange juice is "fresh," marketers have thus become more brazen. They have boldly appropriated the word rather than merely the packaging that once symbolized it. Hence P&G's "Fresh from concentrate" orange juice oxymoron.

As early as 1963 the FDA publicly said no to labeling as "fresh" commercially processed orange juice. In 1969 the agency added teeth to the prohibition by declaring that the use of the word for any heated or chemically processed food would be deemed "false and misleading."[7] In 1991, in answer to the FDA's repeated letters to P&G that the word "fresh" was reserved for producers of raw, unheated juice, P&G spokeswoman Wendy Jacques was resolute: "P&G . . . will stand by the Fresh Choice name." She justified the use of the name because a P&G survey showed that consumers thought "fresh" referred to the juice's taste rather than to its method of production.[8] Freshly squeezed orange juice had become so rare that the 1990s consumer could not imagine "fresh" to mean an actual procedure. If by 1961 freshly squeezed juice had, the juice processor argued during the hearings, become the exception deserving differentiation, by 1991 freshly squeezed juice had, P&G's survey suggested, become nonexistent. The only trace of it was its taste that processors like P&G attempted to imitate.

The juice processor had come a long way since 1961, from trying to convince the FDA that highly processed orange juice should be called "orange juice," to insisting on being permitted to call it "fresh." In 1961 the FDA concluded that pasteurized orange juice processors could not call their product just orange juice; the word "pasteurized" had to be present on the juice carton. Similarly, in 1991 Kessler did not buy P&G's pitch for retaining "fresh choice" in the name of Citrus Hill's from-concentrate. Kessler, like his FDA predecessors, refused to accept that freshly squeezed juice was no more than a memory. It had a shape, not just a taste, which was tangible to the consumer. Labeling Citrus Hill's from-concentrate juice "fresh" was thus, in his opinion, a misnomer.

In 1961 the FDA turned to the standard of identity to stop what it thought was dishonest orange juice advertising. In 1991 the FDA, under Kessler's direction, was more hands-on. At the same gathering where Kessler described the events leading to the Citrus Hill crackdown, he confessed that "the Citrus Hill action had much less to do with freshness than it did with the way we enforce the law." He seized upon the P&G label to set an example. He wanted the food industry to know "we mean business." At the same time he made it known that the decision to target Citrus Hill was not haphazard. All of the FDA's enforcement actions were united, he said, by a fundamental principle: "Simply put, we want consumers to have the information they need to make informed choices—information that is accurate, information that tells the whole story."[9] He acknowledged that federal oversight of food labels was nothing new. Food labels had been regulated at the national level as far back as 1906. The FDA's mandate, in Kessler's words, to "educate the people," *was* new. Kessler spoke in the spirit of the Nutrition Labeling and Education Act, from which the new mandate derived, when he declared: "The role of the FDA is not to tell people what to eat, but to make informed choice possible."[10]

At the 1991 convention of food and drug lawyers, Kessler expressed his hope "to restore the FDA to its former position of preeminence in matters involving food."[11] Reclaiming this position required a departure from former ways. The food standard of identity, which initially allowed the marketer's words to substitute for consumer education by stripping food labels of information regarding the ingredients, was an example of regulation that was no longer appropriate for an agency charged with empowering the consumer.

The FDA's current approach to the standard of identity reflects the change in regulatory mindset that Kessler initiated. Kevin Gaffney, senior research and development manager at Florida's Natural, says that the FDA is no longer in the game of making standards of identity.[12] Similarly, the Tropicana director notes, "The FDA has made it quite clear they have no interest in adding or changing the standards of identity." He speculates that, "rather than changing food standards to change the com-position of food, they've said consumer choice through infor-mation is a better way to go. So there's very little interest on their part to modify the standards."[13] The FDA is evidently develop-ing a reputation for attempting to make "informed choice pos-sible," as Kessler directed it to do more than a decade ago.

A recent controversy between the Florida Department of Citrus and three juice manufacturers illustrates the challenge marketing can pose to facilitating informed consumer choice. In 2004 Minute Maid, Wal-Mart, and Tropicana introduced "light" juice to the market. Calorie-reduced, the new product is the orange juice industry's answer to the twenty-first-century low carbohydrate craze. Rational actors wanting to cut their carbohydrates will not eliminate the vitamin C–rich orange from their diet; they will simply substitute whole fruit for juice. But since such a switch would be bad for orange juice processors and packers, they are appealing to the carbohydrate counter by taking away the "orange" in orange juice. The label of Tropicana's Light'n Healthy orange juice boasts half the calories and sugar of 100 percent orange juice.

Concerned that consumers are mistaking the product for 100 percent juice, the Florida Department of Citrus (FDOC) sponsored a survey of 502 consumers to analyze their ability to differentiate among twenty orange juice products, including light and 100 percent juice. The study found that 37 percent in-correctly believed that all twenty were 100 percent juice. After reviewing these results, the FDOC petitioned Minute Maid, Wal-Mart, and Tropicana, the three major packagers of light orange juice, to redesign their labels. It requested that juice content, which could be found in small script on the side panel with other nutritional information, appear on the front, where statements emphasizing cuts in calorie and carbohydrate con-

tent dominate. The request was not well received. In a letter to
the FDOC, Coca-Cola executive Charles Torrey was adamant
that Minute Maid's light juice labels were not at all "confusing
or misleading." Peter Brace, spokesperson for Tropicana Products, said that Tropicana's Light'n Healthy orange juice package was "clear," containing "facts for people to make informed
decisions."[14] The label Brace was referring to read "Tropicana
ESSENTIALS," "LIGHT 'n Healthy." Smaller type that read,
"Made with pure premium orange juice" separated the two
phrases. Under this the carton read, "1/2 Less Sugar & Calories." The label's bottom third depicted a whole orange, encir-
cled by a tape measure, next to a juicy-looking orange half.

   Although the product does contain "pure premium orange
juice," the percentage of this, which is not apparent, is 42. And
the product does have 50 percent less sugar and fewer calories,
but this is because the drink is mostly water and artificial sweetener, which is used to replace the reduction in natural orange
sugar.

   The FDOC did not agree with Brace that together the
two statements on the Light'n Healthy front label enabled the
consumer to make "informed decisions." It wanted the consumer to know precisely what percent of not only calories and
carbohydrates but also orange juice the processor cut out. In
demanding that the drink carton declare juice content "clearly"
on the front, the FDOC was essentially fighting for the type of
enabling label that the consumer was asking for in the person
of Faith Fenton during the 1961 hearings.[15]

   With no control over juice marketed beyond Florida, the
FDOC looked to the FDA for support, and Tropicana has since
modified its "Light'n Healthy" juice label. The front label no
longer states that the product is "made with pure premium
juice." Instead, the juice is called "LIGHT'N HEALTHY"
"TROPICANA PURE PREMIUM." And it no longer advertises

"1/2 Less Sugar & Calories." But it does not betray the fact that the product is only 42 percent juice. That information remains on the side panel, above the breakdown of the product's "Nutrition Facts." While the FDOC seems to have made some headway in encouraging reduced-calorie juice manufacturers to modify their labels, apparently it has not succeeded in convincing the FDA to force them to disclose all the facts on the front panel.

In 1991 "fresh" was not the only term that Kessler targeted in his campaign against P&G's Citrus Hill orange juice advertising. He also objected to P&G's claim, "We Don't Add Anything." The FDA told P&G the statement was misleading because the juice did have additives. It contained flavor enhancers such as orange oil and essence.[16] Although P&G did not honor the FDA's request to remove "fresh" from the Citrus Hill label, the company did agree to stop saying that it did not add anything.

Fast-forward thirteen years. The FDA's failure to follow through on the concession Kessler won from P&G in 1991 gives reason to question whether the FDA, under new administration, will come through for the "carb–conscious" consumer. Tropicana Pure Premium also contains orange oil, and, as those on the inside say, most likely essence too. Yet at the stoplights leading to Tropicana's Bradenton location drivers could, in 2004, read a billboard announcing: "Nothing added. Almost Nothing Removed."

"We Don't Add Anything." (P&G, 1961)
"Nothing Added." (Tropicana, 2004)

The shrewdest lawyer would be hard pressed to find a distinction between the two. The deletion of the former from circulation only to be replaced by the plastering of the latter at a central intersection highlights the FDA's lack of success regulating orange juice advertising. The juxtaposition of the two slogans, the former long ago silenced, and the latter standing unchallenged thirteen years later, captures the inconsistency in FDA enforcement that Kessler set out to end.

Kessler relied on deterrence to purge the FDA of the incongruous actions that he believed detracted from the agency's credibility. He argued: "There is a compelling case to be made that strong enforcement provides the greatest possible incentive to ensure compliance with the food and drug laws of this Nation." Tropicana's resurrection of "Nothing added" despite P&G's concession to inter the equivalent reveals that Kessler was overly optimistic that the food industry would listen if the FDA spoke strongly. He gave the industry the benefit of the doubt when he concluded that the use of "fresh" on highly processed products demonstrated a lack of understanding on the part of some within the industry that the FDA disapproved. Given the FDA's repeated denunciation of "fresh" in association with processed foods, P&G's appropriation of the word for its reconstituted orange juice signaled indifference rather than ignorance. Tropicana's "nothing added" slogan only reinforces the ability of juice manufacturers to get away with not heeding what the FDA has to say.

Kessler acknowledged that his trust in the capacity of "strong" enforcement to ensure FDA compliance was a product of necessity. He admitted that the case he made for deterrence "is especially true for an agency that, by the real-world limitations on its resources, cannot place an FDA cop on every manufacturing corner."[17] The 2004 Tropicana billboard above 9th and 13th Street East in Bradenton is a sign of the weakness of the power of example. To avoid the stain of inconsistency, the FDA seemingly *does* have to place "a cop on every manufacturing corner."

XV

Pleasing Mrs. Smith

At the 1957 meeting of the Florida State Horticultural Society, Robert Brink of Kroger Food Foundations spotlighted the woman with whom everyone engaged in food production—from producer, to retailer, to regulator—was familiar: "But it is the consumer, our Mrs. Smith, who represents the buying power, and by utilizing this power causes products to grow in sales. Thus for any item to realize its full sales potential, Mrs. Smith must be kept food-happy."[1] The food industry's recognition that Mrs. Smith held the buying power did not necessarily work to her advantage. Industry leaders set out to seduce her with sweet words, regardless of whether the products she bought were all they were made out to be.

In 1954 the USDA dedicated its yearbook to the subject of marketing. It sought to lay down guidelines for an industrial sector that was increasingly relying on advertising to win Mrs. Smith's favor. One article listed four principles that should be obeyed. The first two are timeless and uncontroversial: "(a) Good advertising aims to inform the consumer and help him to buy more intelligently; (b) Good advertising tells the truth, avoiding misstatement of facts as well as possible deception through implication or omission. It makes no claims which cannot be met in full and without further qualification. It uses only testimonials of competent witnesses."[2]

Orange juice advertising, beginning as early as the 1950s, failed to follow basic advertising etiquette. The ads of yesterday and today have consistently shortchanged consumers with respect to information. From-concentrate juice manufacturers advertise their product as fresh. Not-from-concentrate juice manufacturers try their best to cover the fact that their product has been pasteurized. Ads for major brands such as Tropicana and Citrus Hill have been found guilty of misstating the facts and deceiving through "implication and omission," to borrow the words of the USDA yearbook. The orange juice consumer has been left in a state of confusion and so is unable to "buy more intelligently."

Arthur Schlink, co-author of the seminal critique of processed foods, *100,000,000 Guinea Pigs*, was always skeptical of advertising. In his 1935 book *Eat, Drink and Be Wary* he advised:

> The instinct which guides the cat and the robin to a proper selection of food for health and healthful activity, will work for man, and has worked for some hundreds of thousands of years for many races in many climes. If that be granted, it is clear that there is no reason whatever why the factory owner and the advertiser should be permitted to dictate to human beings on any subject so delicate and so important, individually and racially, as the determination of one's choice of food and its quality, purity,

and freedom from adulteration. Let your mind and your instincts (and in civilization, you will need both) determine your diet: let others follow the bill-boards and *The Saturday Evening Post*.[3]

Although Schlink was cynical about advertising's service as a device for information, his alternative is no longer apt. When Schlink was writing, mind and instinct may still have been appropriate guides for the food purchase decision. With the number and complexity of processed foods currently on the market, more is now needed.

IV
Orange Juice in the Twenty-First Century

XVI
Where To?

The introduction of FCOJ after the war placed a middleman, the processor, between the Florida orange grower and North American orange juice consumer, and this group sparked the explosion in advertising that has benefited neither grower nor consumer. Processors and brand labels, in the case of Tropicana one and the same, doubly benefit from their product's incessant promotion: advertising pushes consumers to buy more orange juice; and the resulting increase in demand for oranges puts pressure on supply, which inevitably ends in an orange glut, providing processors a cheap source of raw materials. Headlines that read, "Orange-Price Drop Hurts Growers, Is Boon to Tropicana," are not unusual.[1] Further, they are a constant reminder that the 1940s prediction that FCOJ would save the growers' golden groves was off the mark. Initially touted as a "Cinderella product," FCOJ, and the processed orange juice products that it inspired, did not bring about a fairy-tale ending for either grower or consumer. Despite FDA regulation, the juice that swept growers off their feet in the late 1940s evolved into products that have been promoted to the point of the grower's defeat and the consumer's deceit.

The orange grower and juice consumer have two routes to a better future. The first is to sidestep the processor. Without the processor standing between them, the grower and consumer could speak directly to one another. After years of butting heads, the two could begin a friendly, mutually beneficial conversation. The second route is consumer education.

In the twenty-first-century age of advertising and outsourcing, Florida orange growers must be creative if they want to prevent their groves from being paved over. Rory Martin sees benefits to his orange growing business of removing the processor as a mediator between him and the consumer. He has invested in a modest commercial juice extractor to squeeze his own fresh juice. The same box of oranges that sells for three dollars to Tropicana brings in thirty dollars when he juices the oranges himself or sells them whole to the niche market that he is working to establish. Although he markets only 25 percent of his fruit himself, he hopes to continue to grow demand for his fresh products.[2]

Martin is at the forefront of a movement among growers to expand the fresh-squeezed juice market—a movement that processors are resisting. The growers are inspired by the commercial success of not-from-concentrate orange juice. Jim Griffiths, who is leading this movement, considers NFC's rapid growth in sales as a lesson in consumer preferences. The juice's popularity shows that consumers care about how much their juice is processed. Given NFC's success, Griffiths expects that consumers will respond favorably to a juice that is truly fresh.

Tropicana, which has put a lot of money into telling consumers its NFC is as good as fresh-squeezed, and other processors do not want an authentic fresh-squeezed to take the thunder from, or blow the cover on, the imposter NFC. Griffiths says: "They want their product to be perceived as fresh squeezed," and he says that "they're causing problems at the [FDOC's] Citrus Commission. When we try to do something about fresh squeezed they don't, their reps make snide remarks. . . . They raise scare stories."

One such story concerns the fact that every batch of oranges picked for squeezing inevitably contains some that have been gleaned from the ground. Tropicana argues that unless the juice is processed the risk of bacterial contamination is too high. Griffiths admits that oranges that have fallen to the ground increase the bacterial load of the juice. Yet he insists that marketing a safe, freshly squeezed juice is "perfectly possible," and is confident that "if you squeeze the oranges within a reasonably clean environment and sell the juice within seventy-two to ninety-six hours, and keep it below forty degrees after it's been squeezed, you won't have any problems."[3]

Initially the Florida Department of Citrus was not supportive. Griffiths blames the FDOC's director of fresh fruit marketing at the time, who, he says, "hasn't pushed it the way he ought to." That was before the hurricanes of August and September 2004. While the more sturdy processing plants withstood the high winds, the hurricanes ripped up much of Florida's 2004–2005 citrus crop and fresh fruit packinghouses. The FDOC recognized that it had to act quickly to help the severely damaged growers and fresh fruit sellers. Ignoring the processors' concerns, it launched a "fresh-squeezed citrus juicing program" one month after the hurricanes.

The program is not the first of its kind. In the early 1990s about 60 percent of the fresh oranges that Florida sold were juiced in stores. However, a few non-citrus-related incidents of food-borne pathogen outbreaks raised safety issues that ended in-store fresh orange juice manufacture.[4] The FDOC has since conducted polls indicating that consumers want fresh juice and like watching it squeezed in front of them. Under the new program the FDOC has promised to work with retailers to boost the sales of fresh-squeezed juice by introducing extractors into their stores.

Griffiths did not need the hurricanes of the summer of 2004 to persuade him that reviving interest in unprocessed juice would help Florida's struggling orange growing industry. In the spring before the hurricanes hit he was already pondering the possibilities of commercializing freshly squeezed juice: "We could sell five million boxes more than what we're selling today if we got at it right and did the research to make sure we know the answers to some things we don't have today." Florida's growers and packers of oranges for fresh consumption would benefit the most because this group has been having a difficult time selling their fresh product now that consumers have become accustomed to drinking their oranges. When consumers do buy oranges whole, they are more likely to choose California's thick-skinned, easy-to-peel-and-separate navels, which do not grow well in Florida. The FDOC expects that installing juice extractors in supermarkets will provide retailers an incentive to buy more whole Florida oranges, which, without a market, are left to rot on the tree. Although juice processors are reluctant to allow another competitor into supermarket refrigerators, the FDOC is acting in the interest of the grower *and* consumer, who, surveys say, is thirsting for the return of fresh.

The proponents of the FDOC's fresh-squeezed juice program acknowledge that the program alone cannot cure Florida's ailing orange growing regions. Of his proposal Griffiths says, "Because [freshly squeezed juice] tastes better, you can sell it at a premium price. Packinghouse people need a source and place to sell fresh oranges. The more people that drink fresh-squeezed orange juice and get that full bodied flavor when they do it, the more apt they are to want to buy orange juice of some other kind. And when they get tired of paying the high price of fresh squeezed . . . ," his voice trails off in obvious recognition that consumers will return to processed juice, which will suddenly look like a bargain.[5] Cognizant that juice processors are too embedded in Florida's citrus industry to ignore, he has attempted to fashion a plan that suits everyone.

The Florida citrus industry has come full circle. In the early 1900s it realized the boon that fresh juice consumption would be to a state brimming with orange trees. It delivered juice extractors to homes everywhere, turning fresh-squeezed orange juice into a national treat. Then came FCOJ. Its year-round availability, convenience, and affordability promised a seemingly bottomless glass for Florida oranges. But a processed orange juice that appealed to the consumer exacerbated rather than answered the state's orange agricultural problems. FCOJ founded an orange juice industry that grew out of the grower's control. The state of Florida orange agriculture is back to where it started, on shaky ground. The tremors that today's bulldozers create as they clear Florida's land for real estate can be felt by every grower. Raw juice may be the only way out of the straits that the Florida orange grower is trying to squeeze through.

Those in Florida who care about the grower have taken to the roads with equipment that will reacquaint the nation with the real taste of freshly squeezed orange juice. This time the grower's envoys are not, as they were in the late 1920s, knocking on the doors of consumer homes. They are bringing juicers to the food emporium, which in the twenty-first century has nearly replaced the domestic kitchen as the site of food preparation. In theory the consumer, processor, and Florida orange grower will all benefit. In practice, time will tell.

Increasing the availability of freshly squeezed juice is one way to help Florida orange growers make a living, and to enable orange juice consumers to know what they're drinking. There is another option that requires collaboration among government agencies and grower and consumer groups. Even processors, if willing, could provide constructive input. This involved course of action is educating the consumer.

Revisiting the FDOC's dispute with manufacturers of light orange juice illuminates the link between the educated consumer and the future of Florida citrus. One of the FDOC's chief issues with light orange juice was its recipe, which is short on oranges. The core of the FDOC's concern was that light orange juice would reduce the demand, and thus the price paid, for the Florida orange.[6] Its target of the juice label assumes a connection that is not always made between consumer education and the grower's salvation. It believed that if juice percentage were made more apparent to consumers, they would vote with their dollars for a product made from oranges only. Such a choice would ultimately benefit Florida's orange growers.

If the FDOC's perception of the product label as the cause of and cure for an agricultural dilemma is unique, so was its proposal for the light orange juice label. The label it envisioned, one with juice percentage declared front and center, would depart from tradition. As professor of communication studies Matthew McAllister notes, advertisers are inclined to emphasize the pros of consuming rather than price of producing a product. They tend to exclude information that is linked to the cost of a commodity's mass production, such as the quantity of raw materials used in the making. The effect, he says, is that "because advertisers do not include this information and because advertisers so heavily stress consumption views of commodities over production views, we as users of the advertisements may perceive commodities and products overwhelmingly in terms of consumption."[7] The FDOC's call for light orange juice producers to plainly communicate the percentage of raw materials they use in their juice would, if implemented, re-orient the light orange juice label to enable consumers to see the product in terms of production.

Australia's citrus industry is also taking a production-focused approach to the product label. In January 2006 an organization of Australian citrus growers unveiled a new plan to address the growing difficulty Australian growers are having in competing with imports of orange juice concentrate. It designed a logo, which comprises a green circle with a smiling orange in the middle and the words "supported by Australian citrus growers" written below, so that consumers can identify juice squeezed from Australian oranges. The idea is that, given the information, consumers will choose to buy local.[8]

The FDOC and Australian orange growing industry recognize that consumer awareness of the processed foods they buy impacts the farms where the foods, however distantly removed, have their roots. The same cannot be said of the U.S. government. The interdependency between food consumer and farmer is yet to be reflected in the structure of U.S. federal food regulation. The network of roads that has long separated grower from consumer is reproduced at the government level. Overseeing the grower are the Environmental Protection Agency and United States Department of Agriculture, which monitor agricultural production in the United States. Looking after the consumer at the other end of the federal regulatory map is the Food and Drug Administration, which regulates the food that is finally consumed.

While citrus grove owner Jim Brewer's story about his neighbor highlighted the hard time the grower and consumer are having getting along, buried within the 1961 orange juice hearings is an example of the antipathy that has long colored the relations between the consumer's and the grower's respective chaperones, the FDA and the USDA. During the hearings a peripheral dispute arose over the USDA's refusal to cooperate with the FDA's request to send one of its men, Lamar English, to testify. The USDA's obstinacy seemed especially unreasonable considering that USDA standards of quality for processed orange juice were already on the books. Testimony advanced during the hearings established that the difference, aside from authorship, between a standard of identity and standard of quality is hard to distinguish. In light of the resulting orange juice regulatory overlap, both the FDA and the USDA would have had something to gain by the USDA's attendance at the hearings. The FDA could have avoided duplicating research that the USDA had already performed. It also could have learned from the USDA's experience. At the same time, the USDA could have guided the FDA away from writing standards that would make those of the USDA redundant.

The FDA was prompted to ask the USDA to lend English for the day over the issue of whether the FDA should allow previously frozen juice to be added to pasteurized juice. David Kerr, lawyer for the National Orange Juice Association, represented processors who opposed the use of frozen in pasteurized juice. He called a witness who quoted English, but without English's direct testimony the witness's words were hearsay. To the dismay of everyone except FDA officials, no amount of coaxing persuaded English to appear in person. FDA lawyer Bruce Brennan explained to the bewildered attendees the nature of FDA and USDA relations: "[The FDA and the USDA] are oftentimes contrary to one another. . . . This is a common thing."

Another of Kerr's witnesses, Robert Saltzstein, president of the Dairy Service Corporation and secretary-treasurer of Chilled Citrus Products, spoke about the tension that prevented cooperation between the two agencies: "It is regrettable that this record can't be clarified because of it. It is all one government."[9] When the hearings resumed a couple of days later, the USDA's refusal to have English testify at the hearings was the first item on the agenda. Although Kerr agreed that English should testify, he provided a different reason than Brennan's for why he would not: "The Administration does not wish to embarrass one agency at a hearing conducted by another and will not use [its] good office to make a witness available."[10] Whatever the reason, the incident underscored the inefficiency of having two antagonistic agencies regulating food production.

The federal government's division of responsibility for agriculture, on the one hand, and food processing and consumption, on the other, perpetuates the fallacy that grower and consumer live independently. The diminishing distance in Florida between orange grower and juice purchaser is symbolic of the proximity between the two that geography and government have masked. As grower and consumer move closer, so too must their overseers. The solicitation by the FDOC, an agency that at least in theory represents the grower, of the FDA's assistance in averting what it perceives to be an agricultural disaster—an orange juice product short on oranges—is one step toward bridging the gap between grower and consumer, agriculture and processed food regulator. Such a reunion may be the best hope for keeping a high-quality Florida orange juice on the market.

XVII
Orange Juice Speaks Volumes

Two related issues dominated the 2004 presidential election campaign: outsourcing and homeland security. With respect to the former, talk centered on the loss of jobs in the manufacturing sector to foreign nations. With respect to the latter, discussion focused on the need to become less dependent on other nations and to further develop this nation's energy supplies, meaning oil. Yet despite the severe economic crisis that came to dominate the 2008 presidential election, the two major-party candidates made little more than passing reference to agriculture and its relation to these two issues.

Indeed, little mention has been made of entire crops moving to nations with cheaper labor and less stringent environmental regulations. Similarly, when the discussion turns to importance of becoming more self-reliant with respect to energy, the most fundamental source of energy, food, is rarely discussed. The result is relative silence regarding the role domestic agriculture plays in strengthening national security. The rising cost of food, which has led to a global food crisis, is finally drawing attention to the fact that such a disconnect is counterintuitive. For any nation to be truly secure it must be self-sufficient, and to be self-sufficient it must first and foremost be able to feed itself. Almost two centuries ago Thomas Jefferson advised potential émigré Jean Batiste Say that it is "the patriotic determination of every good citizen to use no foreign article which can be made within ourselves." He encouraged Say, in the event Say decided to immigrate to the United States, to apply himself to the manufacture of cotton and establishment of a farm. Jefferson regretted Americans becoming "manufacturers to a degree incredible to those who do not see it." Returning to the land was to him an important step toward "securing [the United States] against a relapse into foreign dependency."[1]

The flood of food imports into the United States during the last couple of decades is evidence that policymakers and the public are ignoring the relationship between agricultural independence and homeland security. Before World War II fruit imports amounted to about one-seventh of domestic production. The figure dropped to one-tenth in the early 1950s. Bananas—a crop that in Jefferson's words cannot "be made within ourselves"—accounted for most of that figure. Since then, imports are threatening U.S. crops as different as apples and oranges and, centuries after their arrival, as American as apple pie and not-from-concentrate.[2]

The loss to other countries of crops long grown in the United States destabilizes not only the homeland but also homesteads. In Florida, urban development is indiscriminately encroaching on irreplaceable natural and agricultural resources such as Orie Lee's orange grove. The impending loss of these groves is part of what nursery owner Roland Dilley recognizes as a larger trend of depending on "everything we grow to come

from way down south." Fruit growers are not the only producers losing their livelihoods to international competition. Major
crops such as soybeans are moving to South America. "The
whole farming industry in this United States is the pits," says
Dilley. "We can't all work at McDonalds and make sales calls
and that kind of thing. It's terrible."[3]

Moving manufacture, whether industrial or agricultural,
to the lowest-cost producer is free trade dogma. Although the
raw numbers say juice oranges are cheapest to grow in Brazil,
there still is reason to question the free market's cost assessment. Florida remains the most productive orange growing region in the world; acre for acre it yields the most fruit. Robert
Barber, an economist at Florida Citrus Mutual, estimates that
Florida produces 50 percent more pounds per acre of orange
sugar solids, the basis upon which juice oranges are bought
and sold, than its most formidable competitor, Brazil. Florida
grower Jim Brewer contends that central southwest Florida "is
the best citrus growing area in the world . . . because of the climate, because of the situation we're sitting in."[4]

The juice orange's relocation from its historic home in
Florida to Brazil is fueled not by superior agricultural geography but by low wages and minimal environmental regulations.
Although Brazil can grow and process oranges at an unbeat-
able price, the bargain has been negotiated on the backs of
labor and land. Free trade is supposed to bring about production efficiency, but the efficiency it promotes can be costly in
terms of humanity.

The increasing reliance of the United States on distant
food sources for its sustenance has repercussions that extend beyond homeland, homestead, and human and environmental security. The more removed that consumers are from the foods

on which they subsist, the more foreign the wisdom of Wendell Berry's basic observation: "Eating is an agricultural act."[5]

In a letter to Arthur Greeves, C. S. Lewis recounted a conversation with J. R. R. Tolkien that captures the state of the twentieth- and, even more so, the twenty-first-century consumer:

> Tolkien once remarked to me that the feeling about home must have been quite different in the days when a family had fed on the produce of the same few miles of country for six generations, and that perhaps this was why they saw nymphs in the fountains and dryads in the wood—they were not mistaken for there was in a sense a *real* (not metaphorical) connection between them and the countryside. What had been earth and air and later corn, and later still bread, really was in them. We of course who live on a standardised international diet (you may have had Canadian flour, English meat, Scotch oatmeal, African oranges, and Australian wine today) are really artificial beings and have no connection (save in sentiment) with any place on earth. We are synthetic men, uprooted. The strength of the hills is not ours.[6]

Modern science now has the capacity to show what is "really" in us. With a mere hair or fingernail sample biologists can trace the carbon that *is* each of us to specific plant species using a mass spectrometer. Michael Pollan, self-described food detective, is working with Todd Dawson at the University of California–Berkeley to determine precisely how much Americans "are" corn. As it turns out, the percentage is high.[7]

In the days that Tolkien conjured, when "the family fed on the same few miles of country," the corn that came from the surrounding earth and air "really was in them." Not true today. The corn in most of us does not grow from the air we breathe or earth beneath our feet. It travels from sparsely populated places, from faraway fields in Iowa, for instance. From there it is shipped to processing stations across the nation, where it is turned into unrecognizable forms such as cattle feed for meat and high fructose corn syrup for a rainbow of sweets. Whatever shape the corn takes, Pollan's tests say it is sticking to the bones, and, as obesity rates say, not in a good way. It accumulates inside us, an alien substance as plastic as the wrapping of the myriad foods in which it is an ingredient.

The same goes for much of modern foodstuffs. A business executive in New Jersey has no relationship to the orange juice he drinks from Brazil. The connection that he may have "in sentiment" to Florida is due to successful marketing campaigns that have equated orange juice with Florida. The connection is as artificial as the man who is the synthetic foods that he eats.

The "standardised international diet" of which Lewis spoke does not, as the act of eating once did, root Americans in their surroundings. The distance between the soil that grows the foods Americans eat and the concrete ground the majority walks on fosters indifference to the environment. The nymphs and dryads that Tolkien believed animate the world remain invisible so long as food travels as far as it does to reach the consumer's plate. Government and industry characterization into the 1960s of the food consumer as "Mrs. Housewife" is partly to blame for the average consumer's obliviousness to where and how that individual's food is produced. This label carried assumptions that led to the construction of flawed food regulatory structures. The standard of identity is one that failed to work the way it was supposed to regarding the "interest of consumers." For decades it stood as an obstacle to the advancement of knowledge about the multifarious foods contained under its umbrella.

Disseminating the truth about processed orange juice, a product that since its debut has been promoted as Florida freshness concentrated in a glass, has the potential to wake consumers to how damaging food ignorance can be. Consumer unawareness about where and how the oranges in their juice are squeezed has allowed processors to spin tales that hide not only the extent of orange juice processing but also the struggling Florida orange growing economy.

The Florida Department of Citrus is betting that more forthright orange juice labels will encourage consumers to buy products that will benefit the Florida orange grower. It believes consumers want the real thing: 100 percent Florida squeezed. There may be little chance of returning to the days that Tolkien remembers, when families "fed on the produce of the same few miles of country." However, food education promises to bring consumers closer, in spirit if not in body, to the producers of their food. In so doing it has the potential to shake consumers from their complacency and nurture the compassion that is essential to strengthening ailing agricultural sectors in the United States and elsewhere.

XVIII
The Right Fight

In the mid-1940s the Food and Nutrition Board of the National Research Council (NRC) designated oranges, tomatoes, and grapefruit as a food group. The trio appeared second on the NRC's list of the "Basic Seven" food groups, a seminal guide to healthy eating. The orange had officially entered the nutritionist's vocabulary.

In the late 1940s three men invented a saleable frozen concentrated orange juice. It came just in time for Florida growers, who were having trouble getting rid of their surplus oranges, and consumers, who now had a way to get their daily dose of oranges in a form that was economical and always available. Florida, which grew an especially juicy orange, became home to a burgeoning juice industry.

Advertising played the final part in the orange's transformation from a luxury fruit into a perceived life necessity. Beginning in 1948 Bing Crosby crooned for Minute Maid. In the 1960s Anita Bryant brought the Florida Sunshine Tree into households across America. In the same decade the FDA acted on evidence that much of orange juice marketing misinformed the consumer. After convening hearings in 1961, the agency resolved to regulate all forms of orange juice, including the one that Bryant reconstituted.

FDA regulation did not, as juice processors feared, impede the evolution of Florida's orange juice industry. In 1978 the USDA ranked oranges ninth—between potatoes and rice—on its list of the fifteen most valuable crops to the United States.[1] Oranges grown for juice contributed greatly to the orange's strong showing. In the 1980s the convenience of ready-to-serve reconstituted orange juice made obsolete the cans of frozen concentrate that Bryant promoted. In the 1990s "Not From Concentrate" jumped to the top of processed orange juice charts. In the 2002–2003 season, Florida produced more than 203 million bushel-boxes of oranges, almost all of which were squeezed into juice.[2] University of Florida studies show that citrus has generated nine billion dollars of economic activity annually for the state.

Today Florida growers are skeptical that the money will continue to flow. Competition from Brazil, land development pressures, overproduction, and a perennial battle against insects, disease, and weather are all challenging the steadfastness of Florida's Sunshine Tree.

While Florida orange growers worry about their future, consumers continue to drink orange juice with little or no idea about the who, what, where, and why behind their buy. Sophisticated advertisements, combined with misdirected regulation focusing on product content rather than on consumer education, have deceptively created a national habit. Look no further than not-from-concentrate orange juice, the only category of processed orange juice that continued to grow into the twenty-first century. Consumers pay a premium for qualities that the juice does not carry. Marketed as fresh and "unfooled" with, NFC is heat-treated and heavily handled. And contrary to what many consumers believe, much of NFC is not a purely Florida product anymore.

It is difficult to find an orange juice consumer who is not bothered by the fact that a product that is made out to be fresh sits in storage, sometimes for upward of a year, and is made palatable only by the addition of a flavor pack. As Americans we say in the U.S. Constitution and elsewhere that we value individual autonomy and free choice. Yet little is being done about the deceit and ignorance that guide consumers through the maze of modern supermarkets.

The consumer's misconception of NFC, and of commercial orange juice generally, raises the question of the regulatory role of a consumer right to know how food is produced. The concept of a more general community right to know began to take serious shape in the mid-1980s as a result of a deadly gas leak in Bhopal, India, where Union Carbide was operating a plant that was producing the pesticide carbaryl. Thousands died or were seriously injured from the accident. To prevent such an incident from occurring in America, the Emergency Planning and Community Right-to-Know Act (EPCRA) was passed as part of the Superfund Amendments and Reauthorization Act of 1986 (SARA), which amended the Comprehensive Environmental Response, Compensation and Liability Act of 1980 (CERCLA). EPCRA and the other state and federal right-to-know laws currently on the books—such as California's Safe Drinking Water and Toxic Enforcement Act of 1986 (otherwise known as Proposition 65) and, more recently, the Food Allergen Labeling and Consumer Protection Act of 2004 (an amendment to the Federal Food, Drug, and Cosmetic Act)—complement traditional regulation with ways of raising community awareness about the potentially harmful substances that exist in the environment and food supply.

Given the historic link between the community right to know and toxic chemical management, it is not surprising that right-to-know laws to date have tended to be limited to situations where some immediate and quantifiable risk, whether to health, the environment, or the economy, results from ignorance. Some may cite the relative harmlessness to human health of commercial orange juice as justification for why a right to know how it is produced should not figure prominently in its regulation. There are good reasons to place boundaries on rights to know. Requiring industry to publicize information about its products offers competitors the opportunity to use this to their advantage. Heavy disclosure requirements also have the potential to stifle industry innovation. Interpreting a consumer right to know how food is produced as a right to full disclosure would be counterproductive. However, if there is a need to cap the amount and type of information that industry discloses, surely no argument can be made against requiring the information that industry *does* release be truthful.

Moreover, restricting right-to-know laws to those situations in which there is an impending risk to health and safety is problematic not only because of the difficulty of proving a potential harm but also because it ignores the many situations in which the consequences of not knowing are far-reaching, if not ultimately life-threatening. For example, the existence of the organic, kosher, and halal certifications, which enable consumers to choose foods that meet certain standards of production, underscores the importance beyond reasons of health and safety of knowing how food is produced. Being able to choose kosher, or halal, or organic may not be a matter of life and death, but that does not diminish the importance of possessing information that makes it possible for people to keep a kosher, halal, or organic household. Because the certifications

---

let consumers purchase foods that accord with their values and beliefs, they further the objective of personal liberty that Americans hold so dearly.

In 1985 the UN General Assembly passed resolution A/RES/39/248, entitled "Consumer Protection," that offers guidelines on the subject. The resolution's "Guiding Principles" list the "legitimate needs" that the guidelines are intended to meet. As a reminder that legitimate reasons beyond preventing harm exist for informing consumers, one is providing individuals with access to information that lets them make informed choices "according to individual wishes and needs."[3] Providing consumers with adequate information to fulfill personal wishes is, according to the UN, a "legitimate need" that all nations should strive to achieve.

The history of the production and distribution of orange juice and orange juice knowledge emphasizes the mistake in ignoring the right to know in situations in which relatively innocuous products are regulated. The result is a nation of consumers often basing their purchase decisions on false information. The implications of consumer ignorance about commercial orange juice are significant for health, personal autonomy, and agricultural and environmental integrity and make a strong case for recognition of a consumer right to know how food is produced. Unless we as consumers are provided with factual information, we cannot accurately assess what and what not to worry about. We cannot properly rank our priorities. We cannot make meaningful choices regarding the massive number of industrial products on the market.

Just before the 1961 standard of identity hearings, then–FDA commissioner George Larrick declared: "The consumer has a right to know what is in his food."[4] The 1990 Nutrition Labeling and Education Act breathes life into the right by requiring that the nutritional profile of packaged foods be labeled, and by deeming mislabeled and misbranded products adulterated. In 1991 then-FDA commissioner David Kessler acted on the act's definition of adulteration to impound reconstituted orange juice marketed as "fresh." Larrick's early words, the act, and Kessler's action illustrate that the FDA has interpreted the consumer right to know as more than a mechanism for harm prevention. But it has done so sporadically, and while it has recognized the importance of providing consumers with information about what is in their food, it has been less forthcoming in providing information about how food is produced.

Beyond the FDA the federal government has taken some measures to provide consumers with information about how food is produced. The USDA-administered organic certification regulations let consumers choose foods that have not been grown with synthetic pesticides and fertilizers. Congress approved the new Country of Origin Labeling (COOL) as part of the 2002 farm bill, which took full effect in September 2008 and which lets consumers know the source of their beef, pork, lamb, seafood, peanuts, and fresh and frozen fruits and vegetables. Both are steps forward, but more is needed. Congress has provided consumers with the right to know where their oranges come from but not the circumstances under which they were harvested and processed. The USDA has provided consumers with a means to choose foods that have not been grown with synthetic chemicals, but the FDA has not provided a way of telling whether certain other foods contain hidden ingredients such as genetically engineered organisms or flavor packs to make processed foods taste fresh. The logical next step is a comprehensive consumer right to know how food is produced.

The time is ripe to reevaluate the reach of right-to-know

laws. As more and more industries become transnational and mobile, traditional state-centered, top-down approaches toward regulation are proving ineffective. In an age in which information is powerful and plentiful, information-based strategies are obvious and appropriate substitutes. Right-to-know laws, historically used to supplement command and control forms of regulation, now stand as promising alternatives.

A consumer right to know is especially critical in the realm of food production. What to eat is arguably the most personal of decisions. The choice of what to eat determines not only each individual's physical being but also the broader physical and human environment in which each individual, as a member of the human community, lives. Until legislators and the federal agencies that implement the laws acknowledge the value of right-to-know laws as a means to freedom not only *from* harm but also *to* choose, consumers will continue to be unable to make truly autonomous, personhood- and environment-defining, decisions.

The Florida Department of Citrus has recognized the connection between consumer food literacy and land use. Its fight to have juice percentage stated on the front of light orange juice labels underscores the imperative of providing consumers with more, and more visible, information about the production of deceptively straightforward foods. Hopefully, federal lawmakers and agencies will listen and accept this as a right in its own right. A consumer right to know how food is produced is long overdue.

Chronology

1400s   Arabs help spread *Citrus sinensis,* the sweet orange, from its native land in Southeast Asia to Europe.

1560   Spanish explorers bring seeds of the sweet orange to Florida.

1821   Commercial cultivation of the sweet orange begins in Florida after Spain cedes Florida to the United States.

1870s   Samuel Parsons establishes Florida's first orange nursery in Lakeland, Florida.

          The Valencia, one of the two orange varieties upon which Florida's orange juice industry is built, is introduced to Florida via a nursery in the United Kingdom.

1879   H. E. Hamlin discovers a chance seedling in his Glenwood Florida grove that he names the Hamlin, now the most heavily planted orange variety in Florida.

1880s   Budding becomes the propagation technique of choice among commercial citrus growers in Florida.

1909   Florida orange growers join to form the Florida Citrus Exchange to manage the state's citrus surplus. The Exchange's purpose is to increase the demand for Florida oranges.

1915   The first U.S. citrus processing plant opens in Haines

City, Florida, to deal with oranges unfit to be sold fresh. The plant gives birth to Florida's canned orange juice industry.

1924   The Florida Citrus Exchange begins to distribute juice extractors to homes across the country. By the end of 1927 a total of 16,324 juice extractors etched with the Exchange's "SealdSweet" signature had made their way into American kitchens.

1935   The state of Florida establishes the Florida Department of Citrus (FDOC) to represent the interests of Florida citrus growers.

1938   The Food and Drug Administration introduces the standard of identity as a form of food regulation.

1945   The National Research Corporation (NRC) incorporates Florida Foods and completes construction of a plant in Plymouth for the commercial production of frozen concentrated orange juice. The NRC proceeds to enter into a contract with Vacuum Foods Corporation to continue production of frozen concentrate. At the end of the decade Vacuum Foods is renamed Minute Maid.

1948   The United States issues a patent for a method of making a palatable frozen concentrated orange juice. The invention, by three government employees, gives birth to the Florida orange juice industry.
       Bing Crosby begins advertising Minute Maid's frozen orange juice concentrate.
       Citrus Mutual, a voluntary trade association of Florida citrus growers, organizes to maximize grower returns.

1949   Florida markets ten million gallons of frozen orange juice concentrate.

1955   Per capita consumption of chilled, or "pasteurized," orange juice grows sufficiently to become a USDA statistic.

1956   The Florida Department of Citrus begins investigating the potential of using orange essence as a flavoring ingredient for orange juice.

1959   The first recorded orange juice pack containing added orange essence is marketed.

1960   Coca-Cola Co. buys Minute Maid.

1961   The FDA holds hearings, which last six months, into defining standards of identity for orange juice.

        James Redd founds Intercit, the first company in Florida devoted to recovering orange essence for resale to juice processors.

1962   The first major freeze hits Florida's processed orange juice industry. To make up for Florida's orange juice shortfall, members of the Florida industry help set up a juice processing infrastructure in Brazil.

1974   The USDA releases the Swingle Citrumelo rootstock, enabling Florida growers to plant their orange trees more densely.

1977   The number of orange juice processing plants in Florida peaks at fifty-three.

1985   Brazilian surpasses Florida orange production for the first time, establishing Brazil as the world leader in orange production.

        Sales of "Ready to Serve" (RTS) juice overtake sales of frozen concentrate for the first time.

        Tropicana directors meet to discuss how to keep the company's pasteurized orange juice competitive with the now-popular reconstituted orange juice.

1990   The FDA's Nutrition Labeling and Education Act is enacted.

1991   FDA commissioner David Kessler impounds a Proctor and Gamble reconstituted orange juice that is labeled "fresh."

1995   The United States begins importing "Not from Concentrate" orange juice from Brazil.

1998   PepsiCo. buys Tropicana from Seagrams.

2004   The number of orange juice processing plants in Florida drops to eighteen, eight of which are operated by multinational corporations with interests in Brazil.

2005   The Florida Department of Citrus expands its mandate to recognize the interests of not only the grower but also the citrus industry.

2006   The Florida Department of Citrus's Citrus Commission forms a blue-ribbon advisory panel to investigate the uncertainty of the Florida citrus industry's future.

Notes

Chapter One:
The Seeds of Florida's Sunshine Tree

1. Robert Willard Hodgson, "Horticultural Varieties of Citrus," in *The Citrus Industry*, vol. 1, ed. Walter Reuther, Herbert Webber, and Leon Batchelor (Berkeley: University of California, 1967), 441; Pinhas Spiegel-Roy and Eliezer E. Goldschmidt, *Biology of Citrus* (Cambridge: Cambridge University Press, 1996), 7.

2. D. J. Browne, *The Trees of America* (New York: Harper, 1857), 60.

3. John McPhee, *Oranges* (New York: Farrar, Straus and Giroux, 1967), 67.

4. Wayne D. Rasmussen, *Agriculture in the United States: A Documentary History* (New York: Random House, 1975), 3:2732; Browne, *The Trees of America*, 63; Robert R. Krueger, "Fruit: Citrus Fruit," University of California, Riverside, undated paper, 2; Jay Mechling, "Oranges," in *Rooted in America: Foodlore of Popular Fruits and Vegetables*, ed. David Scofield, Gillespie Wilson, and Angus Kress (Knoxville: University of Tennessee Press, 1999), 124.

5. Mechling, "Oranges," 124; C. A. Bass, "Historical Sketch of the D. D. Dummitt Grove at Allenhurst," *Annual Meeting of the Florida State Horticultural Society 39* (1926): 234–235.

6. Hodgson, "Horticultural Varieties of Citrus," 454; Carita Doggett Corse, "The History of Citrus in Florida," in *Stories of Florida* (Federal Writers' Project of the Works Progress Administration, 1938–1939), 4; Hamilton P. Traub and Ralph T. Robinson, "Improvement of Subtropical Fruit Crops: Citrus," in *The Yearbook of Agriculture 1937* (GPO, 1937), 749–824.

7. Traub, "Improvement of Subtropical Fruit Crops: Citrus," 770.

8. Hodgson, "Horticultural Varieties of Citrus," 454; interview with Jim Griffiths, April 12, 2004.

9. Hodgson, "Horticultural Varieties of Citrus," 454.

10. Traub, "Improvement of Subtropical Fruit Crops: Citrus," 766; Hodgson, "Horticultural Varieties of Citrus," 446.

11. Interview with Roland Dilley, April 30, 2004. Dilley, a Florida nursery owner, helped develop the container method. He grew the first "container trees" in Florida in 1978; interview with Abraham, assistant to nursery owner Phillip Rucks, April 19, 2004. Abraham took me through the grafting process; Larry K. Jackson and Frederick S. Davies, *Citrus Growing in Florida* (Gainesville: University Press of Florida, 1999), 4:122, 127–141.

12. Interview with Roland Dilley, April 30, 2004.

13. Interview with Michael Kesinger, bureau chief of the Bureau of Citrus Budwood Registration, March 24, 2004. To prevent the spread of disease the budwood program requires that budwood be certified as disease-free before being inserted into rootstock tree.

14. Michael Kesinger, *Annual Report: July 1, 2002–June 30, 2003: 50 Years of Protection and Productivity* (Winter Haven: Bureau of Citrus Budwood Registration, 2003), 21.

Chapter Two:
The Twentieth-Century Squeeze

1. F. M. O'Byrne, "Citrus Varieties," *Annual Meeting of the Florida State Horticultural Society 33* (1922): 147.

2. Interview with Tropicana Employee #1, February 27, 2004. I interviewed two employees at Tropicana. One no longer works there. Hereinafter I will refer to them as "Tropicana Employee #1" and "Former Tropicana Employee" to preserve their anonymity.

3. O'Byrne, "Citrus Varieties," 147.

4. James T. Hopkins, *Fifty Years of Citrus: The Florida Citrus Exchange 1909–1959* (Gainesville: University of Florida Press, 1960), 1; *The Story of a Pantry Shelf: An Outline History of Grocery Specialties* (New York: Butterick, 1925), 103–104; Thomas Palmer's address to members of the newly formed Florida Citrus Exchange, July 22, 1909, as quoted in Hopkins, *Fifty Years of Citrus: The Florida Citrus Exchange 1909–1959*, 7.

5. Interview with Kevin Gaffney, Senior Research and Development Manager at Florida's Natural, March 18, 2004.

6. "History of Florida Citrus-Timeline," *Florida Grower* (August 2000); *Food Consumption, Prices, and Expenditures, 1909–63* (USDA Economic Research Service, September 1981).

7. California agriculture historian Steven Stoll writes of a similar ap-

proach in California toward growing the state's citrus industry. In the 1920s marketers of the fruit produced by the California Fruit Growers' Exchange initiated campaigns to not only educate consumers about the benefits of oranges but also convince them to eat more than they needed. Steven Stoll, *The Fruits of Natural Advantage* (Berkeley: University of California Press, 1998), 87–88; Hopkins, *Fifty Years of Citrus: The Florida Citrus Exchange 1909–1959*, 25–26.

8. R. T. Will, "Some Phases of the Citrus By-Product Industry in California," *Industrial and Engineering Chemistry* 8, no. 1 (1915): 82.

9. Hopkins, *Fifty Years of Citrus*, 80, 92–93. The Florida Citrus Exchange also sent its extractors to distant lands, including England, Ireland, Scotland, France, Italy, Spain, Japan, Alaska, and Cape Town, South Africa. Ibid., 93.

10. Erna Risch, *The Quartermaster Corps: Organization, Supply, and Services*, vol. 1 (Center of Military History Department of the Army, 1953), 175–178.

11. Letter from Colonel L. S. Ostrander, as cited in Harold W. Thatcher, *The Development of Special Rations for the Army: QMC Historical Studies*, No.6 (QMC Historical Studies, G. A. Petersen, 1944), 10; Capt. V. O. Wodicka, "Preservation of Foodstuffs," *Industrial and Engineering Chemistry*, 35 (January 1943): 15, as cited in Thatcher, *The Development of Special Rations for the Army*, 107.

12. Subsistence Research Lab File 430.02, K Ration, Field Tests, 1941–1942, as quoted in letter from Col. Paul P. Logan OQMG to Col. Isker, November 13, 1941, as reproduced in Thatcher, *The Development of Special Rations for the Army: QMC Historical Studies*, 77–78; Subsistence Research Lab Research Report No. 33–43, May 7, 1943, as cited in Thatcher, *The Development of Special Rations for the Army*, 91; Alissa Hamilton, "World War II's Mobilization of the Science of Food Acceptability: How Ration Palatability Became a Military Research Priority," *Ecology of Food and Nutrition* 42 (2003): 331; Walter Porges, *The Subsistence Research Laboratory* (C.Q.M.D. Historical Studies, May 1, 1943), 76–80; See ibid., 75–76, for the percent breakdown of the K ration's calories and page 80 for the ration's calories and gross weight as of 1943. Porges notes that the breakdown was not entirely satisfactory given the National Research Council's recommendation that a ration's fat calories should not exceed 40 percent.

13. Porges, *The Subsistence Research Laboratory*, 78, 87; Charles Romanus and William Ross, *The Quartermaster Corps: Operations in the War Against Germany* 3 (Center of Military History Department of the Army, 1965), 523.

14. Franklin W. Dove, "Food Acceptability—Its Determination and Evaluation," *Food Technology* 1, no. 1 (1947): 2; Franklin W. Dove, *The Process of Developing Food Acceptance Research* (Quartermaster Food and Container

Institute for the Armed Forces, 1943), 7; Herbert Meiselman and Howard Schutz, "History of Food Acceptance in the U.S. Army," unpublished manuscript, 2; Romanus, *The Quartermaster Corps,* 523.

15. Interview with Thomas B. Mack, May 2004.

16. Mike Compton, "Concentrated Effort," *The Ledger,* April 25, 1993.

17. Interview with Edwin Moore, March 5, 2004.

18. Monica Lewandowski, "Formulating Frozen Concentrate," *Florida Grower* (Mid-August 2000): 44–46.

19. J. A. Murray, as quoted in Lewandowski, "Formulating Frozen Concentrate," 47.

20. *Citrus Summary 2002–2003* (Florida Department of Agricultural and Consumer Services, Agricultural Statistics Service, February 2004), 18.

21. Interview with Edwin Moore, March 5, 2004.

22. Ibid.; Hopkins, *Fifty Years of Citrus,* 196; Compton, "Concentrated Effort."

23. Hopkins, *Fifty Years of Citrus,* 222; Lt. Col. H. C. Keeney, "Army Needs in the Citrus Fruit Field," *Conference on Citrus Processing* (USDA Bureau of Agricultural and Industrial Chemistry, 1951).

24. Compton, "Concentrated Effort"; interview with Sharon B. Garrett, Environmental Specialist, Florida Department of Agriculture, Division of Plant Industry, and citrus grower, March 24, 2004.

25. Hopkins, *Fifty Years of Citrus,* 172–173.

26. L. G. MacDowell, E. L. Moore, and C. D. Atkins, "Frozen Concentrated Orange Juice: Development and Significance," *Florida State Horticultural Society* 75 (1962): 319; Hopkins, *Fifty Years of Citrus,* 222–223.

27. *Canner,* September 27, 1952, as cited in Harvey A. Levenstein, *Paradox of Plenty: A Social History of Eating in Modern America* (New York: Oxford University Press, 1993), 109.

28. Interview with Edwin Moore, March 5, 2004.

29. Hopkins, *Fifty Years of Citrus,* 237, 250.

30. *Food Consumption, Prices, and Expenditures, 1909–63* (USDA Economic Research Service, September 1981); *Citrus Reference Book* (Florida Department of Citrus, Economic and Market Research Department, 2003).

Chapter Three:
The Power of Promotion

1. Hopkins, *Fifty Years of Citrus: The Florida Citrus Exchange 1909– 1959,* 226.

2. *Food Consumption, Prices, and Expenditures, 1909–63.*

3. Roland Marchand, *Advertising the American Dream: Making Way for Modernity, 1920–1940* (Berkeley: University of California Press, 1985), 108–109; Ernest Dichter, "On the Psychology of Radio Commercials," in *Radio Research 1942–43*, ed. P. F. Lazersfeld and F. N. Standton (New York: Duell, Sloan and Pearce 1944), 477, as cited in Marchand, *Advertising the American Dream*, 360.

4. Interview with Edwin Moore, March 5, 2004.

5. Marchand, *Advertising the American Dream*, 359–360.

6. Matthew P. McAllister, *The Commercialization of American Culture* (Thousand Oaks: Sage Publications, 1996), 38.

Chapter Four:
Introducing the FDA Standard of Identity

1. Suzanne White Junod, "The Rise and Fall of Federal Food Standards in the United States: The Case of the Peanut Butter and Jelly Sandwich," *Science, Medicine and Food Policy in the Twentieth Century* (FDA, April 9, 1999): 4.

2. Junod, "The Rise and Fall of Federal Food Standards in the United States," 6, 10; see Richard Merrill and Earl Collier Jr., "'Like Mother Used to Make': An Analysis of FDA Food Standards of Identity," *Columbia Law Review* 74 (1974): 561, for a detailed account of the peanut butter hearings and critique of the administrative costs involved.

3. Junod, "The Rise and Fall of Federal Food Standards in the United States," 8.

4. Merrill and Collier, "'Like Mother Used to Make,'" 609.

5. Ibid., 613–614.

6. Earl L. Butz, "Agribusiness in the Machine Age," *USDA Yearbook 1960* (Washington, 1960), 380–384, as quoted in Rasmussen, *Agriculture in the United States: A Documentary History*, 3409.

7. Edward Hampe and Merle Wittenberg, *The Lifeline of America: Development of the Food Industry* (New York: McGraw-Hill, 1964), 4; Rasmussen, *Agriculture in the United States*, 3409.

8. A. L. Brody and J. B. Lord, eds., *Developing New Food Products for a Changing Marketplace* (Lancaster: Technomic, 2000), 2–3.

9. Hampe, *The Lifeline of America*, 6–7; Richard J. Hooker, *Food and Drink in America: A History* (New York: Bobbs-Merrill, 1981), 355; Judith Putnam, "Major Trends in U.S. Food Supply, 1909–99," *Food Review* 23, no. 1 (2000); http://www.cfsan.FDA.gov/~dms/eafus.html; John C. Kirschman, "Effective Regulatory Approval Process for Food Ingredient Technologies," *Critical Reviews in Food Science and Nutrition* 38, no. 6 (1998); "Natural Ver-

sus Artificial," Perfumer and Flavorist International Flavor Symposium, Newark, New Jersey, 1978.

Chapter Five:
Capturing the Interest of the Orange Juice Consumer

1. 21CFR146.

2. FDA Department of Health, Education and Welfare, "Orange Juice and Orange Juice Products; Definitions and Standards of Identity," Washington, DC, Columbia Reporting Company, 1961, 13:1516–20. I have been unable to locate this twenty-seven volume, 3420-page document since finding it at the Thomas B. Mack Citrus Archives in Lakeland Florida. All references to, and citations of, the document are based on notes and photocopies that I took while visiting the archives during the spring of 2004. I trust that my notes are accurate with respect to quotations and page references, and that any mistakes I may have made are minor.

3. Ibid., 1520.

4. Ibid.

5. Ibid., 1524–1525.

6. Ibid., 1520.

7. Ibid., 18:2202–2203.

8. Ibid., 16:1927.

9. II, Section 27, 3–7, as reproduced in Peter Laslett, ed., *Locke: Two Treatises of Government*, 3d ed. (Cambridge: Cambridge University Press, 1988), 287–288.

10. Jackson, *Citrus Growing in Florida*, 267.

11. "Orange Juice and Orange Juice Products," 13:1554.

12. Ibid., 20:2403–2410.

13. Interview with Allen Morris, April 15, 2004.

14. "Orange Juice and Orange Juice Products," 16:1917.

15. Ibid., 25:3134–3136.

Chapter Six:
Regulating Knowledge

1. "Orange Juice and Orange Juice Products," 6:594, 9:922.

2. Ibid., 15:1844.

3. Ibid., 7:725–726.

4. Ibid., 6:595–597.

5. Reuther, Webber, and Batchelor, ed., *The Citrus Industry,* 1:65; Walton B. Sinclair, *The Orange: Its Biochemistry and Physiology* (University of California, Riverside, Division of Agricultural Sciences, 1961), 18–19; McPhee, *Oranges,* 7.

6. "Orange Juice and Orange Juice Products," 6:600.

7. Ibid., 16:1957.

8. Ibid., 7:709L, 710.

9. Ibid., 12:1421.

10. Ibid., 10:1048.

11. Ibid., 17: 2038; letter from Louis G. MacDowell to the Dairy Service, April 29, 1955, as quoted in Ibid., 2034–2039.

12. Ibid., 9:955, 972.

13. Ibid., 982.

14. Ibid., 6:601.

15. Ibid., 13:1520.

16. Ibid., 11:1260.

17. Interview with Edwin Moore, March 5, 2004.

18. "Orange Juice and Orange Juice Products," 9:929, 931, 937, 995.

19. Marchand, *Advertising the American Dream,* 342; Kathryn Weibel, *Mirror, Mirror: Images of Women Reflected in Popular Culture* (Garden City, NY: Doubleday/Anchor, 1977), 167, as cited in Marchand, *Advertising the American Dream,* 351.

Chapter Seven:
Regulating Misleading Orange Juice Labeling

1. "Orange Juice and Orange Juice Products," 13:1536–1537.

2. Ibid., 6:595–597.

3. Ibid., 17:2176.

4. Ibid., 18:2165.

5. Ibid., 8:837.

6. Ibid., 837.

7. Ibid., 27:3400.

8. Ibid., 14:1610.

9. Ibid., 12:1324.

10. Ibid., 11:1288.

11. Fla. Stat. Section 601.19.

12. David A. Kessler, "Remarks Before the National Food Editors and Writers' Association," Washington, D.C., October 3, 1991.

13. "Orange Juice and Orange Juice Products," 11:1288.

14. Tropicana Employee #1, e-mail message to author, December 2, 2004.

15. "Orange Juice and Orange Juice Products," 11:1290–1291.

16. Ibid., 12:1319, 1326.

17. Ibid., 12:1333–1336.

Chapter Eight:
Regulating Content

1. "Orange Juice and Orange Juice Products," 5:561–562.

2. Ibid., 558–564.

3. Ibid., 21: 2458–2460.

4. Ibid., 11:1255–1256, 22:2653.

5. Interview with Tropicana Employee #1, April 29, 2004; 21CFR146.140.

6. "Orange Juice and Orange Juice Products," 24:2970–2972.

7. Ibid., 3000.

8. Ibid.

9. Ibid., 3000–3002.

10. Ibid., 17:2010.

11. Ibid., 2034.

12. Ibid., 24:3016–3017.

13. Ibid.,18:2201, 2205.

14. Ibid., 27:3409–3410.

15. Interview with Renée Goodrich, March 5, 2004.

16. "Orange Juice and Orange Juice Products," 2:174, 195, 212.

17. Ibid., 119; ibid., 4:373.

18. Ibid., 2:211.

19. Ibid., 22: 2673–2677, 2679–2680.

20. Ibid., 2689–2690.

21. CFR146.146.

22. "Orange Juice and Orange Juice Products," 26: 3311–3312.

Chapter Nine:
Regulating the Essence of Orange Juice

1. "Orange Juice and Orange Juice Products," 19:2326.

2. Ibid., 25:3152, 3154.

3. Interview with Firmenich Employee, March 31, 2004. I interviewed one citrus flavorist at Firmenich. I will refer to him as "Firmenich Employee" to preserve his anonymity.

4. "Orange Juice and Orange Juice Products," 8:794.

5. Interview with Renée Goodrich, March 5, 2004.

6. "Orange Juice and Orange Juice Products," 25:3165–3167, 26:3218.

7. Interview with Robert Braddock, professor of Food Science, University of Florida, April 27, 2004.

8. "Orange Juice and Orange Juice Products," 25:3170–3171.

9. Ibid., 26:3294–3295.

10. Ibid., 25:3157–3160, 3173.

11. Ibid., 26:3205–3211.

12. Ibid., 2:103–104.

13. Ibid., 26:3212.

14. Ibid., 3308.

15. Ibid., 3278–3280.

16. Interview with Renée Goodrich, March 5, 2004.

17. "Orange Juice and Orange Juice Products," 2:184–185.

18. Ibid., 25:3174, 3177.

19. Ibid., 26:3210.

20. Ibid., 25:3232.

21. Ibid., 16:1889, 1919–1920.

22. Ibid., 22:2605–2606.

23. Ibid., 25:3157.

24. Ibid., 3177, 3263–3265.

25. Ibid., 26:3196.

26. Ibid., 21:2604.

27. Interview with Tropicana Employee #1, April 29, 2004.

28. "Orange Juice and Orange Juice Products," 26:3200.

29. Ibid., 16:1888, 1918.

30. Ibid., 26:3139.

31. Ibid., 3200–3201, 3204–3205.

32. Ibid., 22:2605.

33. Ibid., 15:1885.

34. Ibid., 16:1889.

35. Ibid., 17:2001–2002.

36. Ibid., 25:3161, 17:2002.

37. F. R. 8511; 22 F. R. 3893; 25 F. R. 1770.

38. F. R. 10900–10903.

Chapter Ten:
Processed Orange Juice Hits Florida

1. Interview with Jim Griffiths, Managing Director of Citrus Grower Associates, April 12, 2004.

2. Stoll, *The Fruits of Natural Advantage,* 88.

3. Interview with Jim Griffiths, April 12, 2004.

4. Interview with James D. Brewer, March 10, 2004.

5. Interview with Robert Behr, March 18, 2004; interview with Jim Griffiths, April 12, 2004.

6. Ronald Muraro, Fritz Roka, and Robert Rouse, *Budgeting Costs and Returns for Southwest Florida Citrus Production, 2004–2005* (Gainesville: Institute of Food and Agricultural Sciences, 2005), 13; Ronald Muraro and W. C. Oswalt, *Budgeting Costs and Returns for Central Florida Citrus Production, 2004–2005* (Gainesville: Institute of Food and Agricultural Sciences, 2005), 8.

7. Interview with Roland Dilley, April 30, 2004.

8. *Citrus Reference Book 2003,* 7.

9. Letter to author from Charles Hendrix, November 11, 2004; Geraldo Hasse, ed., *The Orange: A Brazilian Adventure 1500–1987* (Brazil: Duprat and Lobe, 1987), 165–166; *Citrus Reference Book 2003,* 7, 24.

10. Interview with Jim Griffiths, April 12, 2004.

11. Sergio Barros, *Brazil Citrus Annual 2003* (Global Agriculture Information Network, 2003), 10; interview with Jim Brewer, March 10, 2004; interview with Florida Citrus Mutual Economist Robert Barber, February 26, 2004; Kevin Bouffard, "Citrus Growers to Watch Trade Pact Situation: Less Cash for Tariff Fight," *The Ledger,* November 14, 2004.

12. Interview with Tropicana Employee #1, February 27, 2004.

13. Interview with Lisa Rath, vice president of Florida Citrus Processors' Association, March 15, 2004; Cynthia Barnett, "Does Big Citrus Have a Future in Florida?" *Florida Trend* (March 2003).

14. Allen Morris, "The Economics of Marketing, Advertising and Sales" (lecture, Citrus Research and Education Center, Lake Alfred, Florida, April 6, 2004); interview with Allen Morris, April 15, 2004.

15. Interview with Allen Morris, April 15, 2004.

16. *Citrus Reference Book,* 62.

17. Interview with Tropicana Employee #1, February 27, 2004.

18. Florida Department of Citrus, *U.S. Orange-Juice Imports,* Economic and Market Research Report No. IM-07–12: http://www.floridajuice.com/user_upload/files/ojim1207_47b9deb11d438.pdf; Tropicana Employee #1, e-mail message to author, April 25, 2005; U.S. Department of Commerce,

Imports of Orange Juice, Including Vitamin Added, to the United States (November 2003); interview with Mark Brown, senior research economist at the Florida Department of Citrus, and professor in the Economic and Market Research Department, University of Florida, April 1, 2004.

19. "Plant Report Is Denied," *The Ledger,* February 5, 2005; Kevin Bouffard, "Citrovita Local," *The Ledger,* February 4, 2005.

20. Interview with Roland Dilley, April 30, 2004; Edward Smoak, as quoted in Barnett, "Does Big Citrus Have a Future in Florida?" 1.

21. Interview with Tropicana Employee #1, February 27, 2004.

22. Michael Kesinger, *Annual Report July 1, 2002–June 30, 2003: 50 Years of Protection and Productivity* (Winter Haven: Bureau of Citrus Budwood Registration, 2003), 21; D. P. H. Tucker, S. H. Futch, F. G. Gmitter, and M. C. Kesinger, Florida Citrus Varieties (University of Florida, Institute of Food and Agricultural Sciences, 1998), 13; Bill Castle, "Early-Maturing Sweet Oranges," *Citrus Industry* (February 2003), 19; interview with Citrus Grower Orie Lee, April 20, 2004.

23. Interview with Orie Lee, April 20, 2004; interview with Jude Grosser, professor of horticulture, University of Florida, April 23, 2004.

24. Andy Lavigne, as quoted in Barnett, "Does Big Citrus Have a Future in Florida?" 2.

25. Interview with Orie Lee, April 20, 2004.

26. Barnett, "Does Big Citrus Have a Future in Florida?" 2.

27. Interview with orange grower Jim Brewer, March 10, 2004.

28. Interview with Sharon Garret, environmental specialist, Florida Department of Agriculture, Division of Plant Industry, March 24, 2004.

29. Interview with citrus grower Robbie Martin, February 2004.

30. Interview with Allen Morris, April 15, 2004.

31. Ibid.

32. Florida Department of Citrus, "Florida Department of Citrus: Who We Are, and What We Do!" (2004).

33. Kevin Bouffard, "Citrus Commission Adopts New Statement to Reflect Change," *The Ledger,* April 21, 2005.

34. Florida Department of Citrus, "Florida Department of Citrus: Who We Are, and What We Do!"

35. Interview with Jim Griffiths, April 12, 2004; Jim Griffiths, as quoted in Bouffard, "Citrus Commission Adopts New Statement to Reflect Change."

36. Bouffard, "Citrus Commission Adopts New Statement to Reflect Change."

37. Reuther, Webber, and Batchelor, eds., *The Citrus Industry,* 65; Sinclair, *The Orange: Its Biochemistry and Physiology,* 18–19; Economic Research

Service, *Food Consumption, Prices, and Expenditures, 1909–63;* Economic Research Service, *Food Consumption, Prices, and Expenditures, 1960–80* (USDA, September 1981).

38. Kevin Bouffard, "Panel Will Advise State's Troubled Citrus Industry," *The Ledger,* January 19, 2006.

Chapter Eleven:
NFC Orange Juice Pours into the Nation

1. 21CFR146.140(d)(1).

2. AC Nielsen, retail orange juice sales, in *Citrus Reference Book 2003,* 61–63.

3. Interview with Allen Morris, April 15, 2004.

4. Allen Morris, e-mail message to author, September 4, 2004.

5. Robert C. Evans, "Trends in the Processing of Florida Citrus Fruits and Their Influence on Returns to Producers," *Florida State Horticultural Society Proceedings* 57 (Winter Haven: The Society Office of Publication, 1944), 36.

6. Interview with Daniel King, director of technical services, Florida Citrus Processors' Association, March 15, 2004; interview with Jim Griffiths, April 12, 2004.

7. Interview with Allen Morris, April 6, 2004.

8. AC Nielsen, retail orange juice sales, in *Citrus Reference Book 2003,* 61; interview with Tropicana Employee #1, April 29, 2004.

9. Interview with Allen Morris, April 6, 2004; *Citrus Reference Book 2003,* 61.

10. Interview with Jim Griffiths, April 12, 2004.

11. Interview with Mr. and Mrs. Lee, April 20, 2004.

12. Interview with Allen Morris, April 6, 2004.

Chapter Twelve:
The Orange Juice Wars

1. Interview with Tropicana Employee #1, February 27, 2004.

2. Kevin Bouffard, "Growth, Maid Simple," *The Ledger,* April 5, 2005; interview with Ronald Muraro, extension economist, Florida Department of Citrus, March 4, 2004.

3. Interview with Ronald Muraro, March 4, 2004.

4. Dana Sanchez, "Juice Maker Wants to Nearly Double Its Storage Capacity," *Bradenton Herald,* April 15, 2005; Tim McCann, "Tropicana Wins OK to Build More Storage Tanks," *Bradenton Herald,* May 12, 2005.

5. Interview with Tropicana Employee #1, February 27 2004; 21CR146 .140.

6. Interview with Renée Goodrich, March 5, 2004.

7. Tropicana plant tour guide, April 29, 2004.

8. Interview with Orie Lee, April 20, 2004.

9. Tropicana Employee #1, e-mail message to author, May 12, 2005; *The Orange Book* (Lund: Tetra Pak, 1998), 66.

10. "Orange Juice and Orange Juice Products: Definitions and Standards of Identity," 23:2711.

11. Interview with Allen Morris, April 15, 2004.

12. Tropicana Employee #1, e-mail message to author, May 12, 2005.

13. Interview with Allen Morris, April 15, 2004; interview with Tropicana Employee #1, February 27, 2004.

14. http://www.fmctechnologies.com/FoodTech/FruitsandVegetables/ CitrusProcessing/AsepticCitrusSystems.aspx.

15. Interview with Renée Goodrich, March 5, 2004; interview with Former Tropicana Employee, May 10, 2004.

16. Interview with Tropicana Employee #1, February 27, 2004; interview with Former Tropicana Employee, May 10, 2004.

17. Interview with Renée Goodrich, March 5, 2004; interview with Florida's Natural Employee, March 18, 2004.

18. Interview with Allen Morris, April 15, 2004.

19. Interview with Daniel King, March 15, 2004; interview with Tropicana Employee #1, February 27 and April 29, 2004.

20. Interview with William Castle, professor of horticulture, University of Florida, April 23, 2004; *Florida Agricultural Statistics: Commercial Citrus Inventory 2002* (Florida Department of Agricultural and Consumer Services, December 2002), vii; interview with Frederick Gmitter Jr., professor of horticulture, University of Florida, April 23, 2004.

21. Thomas B. Mack, *Citrifacts II: A Portion of Florida Citrus History* (Lakeland: Associated Publications, 1998), 2:19–20.

22. James Saunt, *Citrus Varieties of the World,* 2nd ed. (Norwich: Sinclair, 2000), 156; interview with Tropicana Employee #1, February 27, 2004.

23. Interview with Rory Martin, February 2004; interview with Lavern Timmer, professor of plant pathology, University of Florida, March 4, 2004.

24. Interview with Jim Brewer, March 10, 2004.

25. Interview with Jim Griffiths, April 12, 2004.

26. Interview with University of Florida professor, whose identity shall remain anonymous, May 5, 2004.

27. Interview with Jim Griffiths, April 12, 2004.

28. Interview with Roland Dilley, April 30, 2004.

29. *American Magazine,* October 1929, 75, as reproduced in Marchand, *Advertising the American Dream,* 340; Marchand, *Advertising the American Dream,* 60, 340.

Chapter Thirteen: Fabricating Fresh

1. Interview with Renée Goodrich, March 5, 2004; interview with Firmenich Employee, March 31, 2004; letter to author from Charles Hendrix, retired chemist for Firmenich and the USDA, November 11, 2004.

2. Interview with Renée Goodrich, March 5, 2004.

3. Interview with Former Tropicana Employee, May 10, 2004; interview with Tropicana Employee #1, April 29, 2004.

4. Interview with Daniel King, March 15, 2004.

5. Interview with Allen Morris, April 15, 2004; interview with Former Tropicana Employee, May 10, 2004.

6. Interview with Firmenich Employee, March 31, 2004; interview with Renée Goodrich, March 5, 2004; interview with Former Tropicana Employee, May 10, 2004.

7. Interview with Russell Rouseff, professor of food chemistry, University of Florida, March 1, 2004; interview with Firmenich Employee, March 31, 2004; interview with Former Tropicana Employee, May 10, 2004.

8. Interview with Jim Griffiths, April 12, 2004; interview with Tropicana Employee #1, April 29, 2004.

9. Interview with Russell Rouseff, March 1, 2004.

10. CFR146.146(a).

11. CFR146.140(a); interview with Daniel King, March 15, 2004.

12. Interview with Kevin Gaffney, March 18, 2004; interview with Daniel King, March 15, 2004; interview with Tropicana Employee #1, April 29, 2004.

13. Interview with Former Tropicana Employee, May 10, 2004; "Orange Juice and Orange Juice Products: Definitions and Standards of Identity," 26:3139.

14. Interview with Firmenich Employee, March 31, 2004.

15. Interview with Daniel King, March 15, 2004.

16. Ibid.

17. Catalina Ferre-Hockensmith (FDA staff member, the Center for Food Safety and Applied Nutrition), e-mail message to author, July 23, 2005.

18. Interview with Tropicana Employee #1, February 27 and April 29, 2004.

19. Interview with Tropicana Employee #1, April 29, 2004.

20. Interview with Renée Goodrich, March 5, 2004; interview with Daniel King, March 15, 2004; interview with Michael Sparks, May 12, 2004.

21. Catalina Ferre-Hockensmith, e-mail message to author, July 23, 2005.

22. Interview with Renée Goodrich, March 5, 2004.

23. Interview with Firmenich Employee, March 31, 2004.

24. Ibid.

25. Catalina Ferre-Hockensmith, e-mail message to author, July 23, 2005. One reason the FDA does not require the labeling of the individual ingredients of a particular flavor, whether natural or artificial, may be that trade secret law generally protects flavor formulas from disclosure.

26. CFR, part 101, Food Labeling, Declaration of Ingredients, Final Rule, 58 Fed.Reg. 2850 (January 6, 1993).

27. "Orange Juice and Orange Juice Products: Definitions and Standards of Identity," 17:2001–2002.

28. Fed.Reg. 2850.

29. "Orange Juice and Orange Juice Products: Definitions and Standards of Identity," 15:1885.

30. Fed.Reg. 2850.

31. Interview with Daniel King, March 15, 2004.

32. Interview with Firmenich Employee, March 31, 2004.

33. Interview with Tropicana Employee #1, April 29, 2004.

34. Interview with Renée Goodrich, March 5, 2004.

35. Fed.Reg. 2850.

36. Interview with Firmenich Employee, March 31, 2004.

Chapter Fourteen:
Moving Beyond the Standard of Identity

1. David Kessler, as quoted in James Bovard, "Double-Crossing to Safety," *The American Spectator* (January 1995): 24–29.

2. David Kessler, "Remarks Before the National Food Editors and Writers' Association," October 3, 1991.

3. Bovard, "Double-Crossing to Safety"; Warren Leary, "Company

Agrees to Drop 'Fresh' from Name of Its Orange Juice," *New York Times,* April 27, 1991.

4. Kessler, "Remarks Before the National Food Editors and Writers' Association."

5. "Orange Juice and Orange Juice Products: Definitions and Standards of Identity," 12:1334–1335.

6. "Orange Juice and Orange Juice Products: Definitions and Standards of Identity," 17:2176.

7. Leary, "Company Agrees to Drop 'Fresh' from Name of Its Orange Juice."

8. David E. Kalish, "Critics Say Companies Avoiding Label Law," *The Ledger,* February 13, 1991.

9. Kessler, "Remarks Before the National Food Editors and Writers' Association."

10. Ibid.

11. David Kessler, as quoted in Bovard, "Double-Crossing to Safety."

12. Interview with Kevin Gaffney, March 18, 2004.

13. Interview with Tropicana Employee #1, April 29, 2004.

14. Kevin Bouffard, "2 Orange Juice Makers Refuse to Change Labels," *The Ledger,* December 15, 2004; Kevin Bouffard, "Citrus Officials Vote to Seek FDA Action on Labels," *The Ledger,* December 16, 2004; letter from Charles Torrey to the Florida Department of Citrus, December 10, 2004, as quoted in Bouffard, "2 Orange Juice Makers Refuse to Change Labels"; Brace's argument echoes former Tropicana vice president David Hamrick's defense of Tropicana's labels during the 1961 hearings (see Chapter 7).

15. Bouffard, "2 Orange Juice Makers Refuse to Change Labels."

16. Kalish, "Critics Say Companies Avoiding Label Law"; Leary, "Company Agrees to Drop 'Fresh' from Name of Its Orange Juice."

17. Kessler, "Remarks Before the National Food Editors and Writers' Association."

Chapter Fifteen:
Pleasing Mrs. Smith

1. Robert Brink, "Mrs. Smith Goes Shopping," *Florida State Horticultural Society Proceedings,* vol. 70 (Winter Haven: The Society Office of Publication, 1957), 260.

2. William B. Ward, "To Sell Goods and Services," *The Yearbook of Agriculture 1954* (GPO, 1954), 181.

3. F. J. Schlink, *Eat, Drink and Be Wary* (New York: Covici Friede, 1935).

Chapter Sixteen:
Where To?

1. Claudia Carpenter, "Orange-Price Drop Hurts Growers, Is Boon to Tropicana," *Bloomberg News,* March 8, 2004.

2. Interview with Rory Martin, February 2004.

3. Garrett Youngblood, "Freezing Out OJ Concentrate," *The Ledger,* April 28, 1996; Mike Compton, "Concentrate Falling Victim to Changing U.S. Lifestyles," *The Ledger,* April 25, 1993; interview with Jim Griffiths, April 12, 2004.

4. Christina DiMartino, "FDOC Rolls Out Fresh-Squeezed Citrus Juicing Program," *Produce News,* October 26, 2004.

5. Interview with Jim Griffiths, April 12, 2004.

6. Kevin Bouffard, "2 Orange Juice Makers Refuse to Change Labels."

7. McAllister, *The Commercialization of American Culture,* 61.

8. Australian Broadcasting Corporation, *New Branding to Clarify Orange Juice Origins,* January 18, 2006, http://www.abc.net.au/news/stories/2006/01/18/1550015.htm.

9. "Orange Juice and Orange Juice Products: Definitions and Standards of Identity," 21:2553–2554.

10. "Orange Juice and Orange Juice Products: Definitions and Standards of Identity," 22:2576.

Chapter Seventeen:
Orange Juice Speaks Volumes

1. Thomas Jefferson to Jean Batiste Say, March 2, 1815, as reproduced in Wayne Rasmussen, *Agriculture in the United States: A Documentary History,* 3:303.

2. Reginald Royston and Arthur Browne, "An Atlas: Fruits and Vegetables," in *The Yearbook of Agriculture 1954* (GPO, 1954), 432; the U.S. Apple Association lists "unfairly priced apple juice concentrate imports" as the first in a series of challenges facing the apple industry (Malinda Miller, *Apple Industry Profile* [Agricultural Marketing Resource Center, Iowa State University, May 2004], 7); many articles have been published regarding the threat that orange juice imports pose to Florida's orange industry (see: Ray Goldberg and Hal Hogan, *Can Florida Orange Growers Survive Globalization?* [Harvard Business School, 2003]); Leo Polopolus and Dan Gunter, "Florida Orange Juice: Natural Disasters, Competitive Market Forces, and the Florida Orange Juice Industry," *Citrus Industry* 66 [1985]; and Cynthia Bar-

nett, "Does Big Citrus Have a Future in Florida?" *Florida Trend* [March 2003]).

3. G. W. Auburn, C. P. Rosson, and C. O. Nyankori, "Soybean Production and Trade Policy Changes in Argentina and Brazil: Implications for the Competitive Position of the United States," *Agribusiness* 7, 5 (1991): 489–490; interview with Roland Dilley, April 30, 2004.

4. Interview with Robert Barber, February 26, 2004; interview with Jim Brewer, March 10, 2004.

5. Wendell Berry, "The Pleasures of Eating," in *What Are People For?* (New York: North Point, 1990).

6. Letter from C. S. Lewis to Arthur Greeves, June 15, 1930, in Walter Hooper, ed., *They Stand Together: The Letters of C. S. Lewis to Arthur Greeves (1914–1963)* (New York: MacMillan, 1979), 363–364.

7. http://www.newfarm.org/features/2005/0805/pollen/index.shtml; Michael Pollan, *The Omnivore's Dilemma: A Natural History of Four Meals* (New York: Penguin, 2006).

Chapter 18:
The Right Fight

1. USDA, *Agricultural Statistics* (GPO, 1978), as cited in Jack R. Harlan, "Gene Centers and Gene Utilization in American Agriculture," in *Plant Genetic Resources: A Conservation Imperative*, ed. Christopher Yeatman, David Kafton, and Garrison Wilkes (Boulder: Westview, 1984), 112.

2. http://www.nass.usda.gov/Statistics_by_State/Florida/index.asp#.html; Michael Sparks and Ismail Mohamed, "Use of Health and Nutritional Information in Marketing Florida Citrus Products," *Ninth Congress of the International Society of Citriculture*, 2000, 1068.

3. http://www.un.org/documents/ga/res/39/a39r248.htm.

4. FDA commissioner George Larrick, as quoted by Faith Fenton in "Orange Juice and Orange Juice Products: Definitions and Standards of Identity," 13:1520.

Index

AC Nielsen, 131, 133
additives, debate over, 91–92
advertising, 181–82, 184; effect of, on FCOJ demand, 108; to increase consumption of oranges, 13, 216–17n. 7; leading to more urgent marketing campaigns, 107–8; stressing consumption over production, 194–95
Africa, 4, 5, 201
agriculture: role of, in national security, 198–99; third great revolution in, 34
American Can Company, 23
American Home Economics Association, 39
American Molasses Company, xxi
apple juice, 141; imports of, 231n. 2
apple pie, American as, 199
apples, 199
aqueous essence, 87
Arabia, 5
artificial flavoring, 96–97
"artificially" (interpretation of), 171
ascorbic acid, 96–97
aseptic storage, 115, 141–44, 169

Atkins, Cedric, D., 19, 21, 22
Auburndale (Florida), 136
Australia, taking production-focused approach to citrus product labels, 195

bananas, 199
Barber, Robert, 200
battery acid, taste of canned juice, 18
Beacham, Lowrie, 42–47, 52, 57, 77, 83–84, 96–97, 99, 101, 102, 166, 168
Behr, Robert, 109
Berry, Wendell, 201
bitter orange, origins of, 5
Brace, Peter, 116, 180
Bradenton (Florida), 69, 137, 139, 140, 141, 150, 182
Brazil: developing vessels to handle NFC shipping, 114–15; as large supplier of orange oil, 165; large supply of land for growing juice oranges, 117; NFC imports from, 115, 116–17, 214; orange juice processors from, influence on

Goodrich, Renée, 80, 88, 94, 138–39,
143, 153, 162, 170
grafting, 8, 9, 11, 145, 216n. 11. *See
also* budding
Griffiths, Jim, 7, 107, 108–9, 122,
123, 129, 133, 147–48, 157, 190–93
Grosser, Jude, 118
groves: in Brazil, 111; diminishing
value of in Florida, 117; distinguished from orchards, 145–46;
growing size of, 144–45; as mono-
cultures, 145; moving south, 109;
Orie Lee's experimental citrus
variety collection, 117–18; turnover of, 149; twenty-year, 147–48
growers: changing relationship with
Tropicana, 120; no longer only
focus of FDOC, 121; relationship
with nongrowers in Florida, 119,
196; representation on Florida's
Citrus Commission, 121
gum arabic, 98
gunpowder, nitrogen used in, 34

halal certification, 207–8
Hamlin, better as juice orange, 11;
discovery of, 7–8; receiving little
attention originally, 11–12; un-
contested popularity of, 9–10
Hamlin, H. E., 7–8, 211
Hamrick, David, 58, 66–71, 74–75,
175–76
hand reaming, 139
Hart, E. H., 6
Hart's Late, 6
Hart's Tardiff, 6, 7
Hendrix, Charlie, 110
Highway 17, 118
Hodges, James, 79
Hodgson, Robert, 6–7

home economists, evolution of,
as profession, 63
homeland security, 198, 199
Homosassa, 11, 12
Hood Dairies, 57
Hooks, Homer, 86
Hopkins, James, 14, 21–22
horticulturalists, 9
hot-pack process, 20
housewives, 173, 202: and interpreting opinions about orange
juice, 51–54, 56–57; presump-
tions about, 59–61; rights of,
42–45
humans, tracing carbon makeup
of, to plant species, 201

identity items, 73–74
imitation orange essence, 96–97
incidental additives, 167. *See also*
processing aids
Indian River: citrus label, 5; location of Tropicana juice processing plant, 67–68; oranges from,
used in Tropicana orange juice,
68–69
industry, rights of, 45–46
innovation, unmanaged, dangers
of, 46–47
in-store fresh juice manufacturing,
191, 192–94
Intercit, 153, 213
international competition, in agriculture generally, 199–200
Islamic forces, 5
Italy, 5

Jacques, Wendy, 177
jams and jellies, standardization
of, 32